Springer Praxis Books

More information about this series at http://www.springer.com/series/4097

David Shonting • Cathy Ezrailson

Chicxulub: The Impact and Tsunami

The Story of the Largest Known Asteroid to Hit the Earth

 Springer

David Shonting
Retired Physical Oceanographer
Adjunct Professor, University of Rhode Island
Naples, FL, USA

Cathy Ezrailson
University of South Dakota
Associate Professor
Vermillion, SD, USA

Springer Praxis Books
ISBN 978-3-319-81897-9   ISBN 978-3-319-39487-9   (eBook)
DOI 10.1007/978-3-319-39487-9

Printed on acid-free paper

This Springer imprint is published by Springer Nature
The registered company is Springer International Publishing AG Switzerland

Preface: About the Book

The story of Chicxulub describes what was likely the most traumatic astrophysical event visited upon the Earth, during the last 100 million years: that of the giant 10-km asteroid that plummeted into the ancient Gulf of Mexico.[1] Moreover, this singular event imposed a far greater disruption to the Earth's biosphere than the strongest earthquake or most violent volcanic eruption. This asteroid's mass (likely half of that of Mount Everest) blasted through the atmosphere within seconds. Its impact at the sea surface set off an explosion that penetrated through a kilometer or so of seawater and into the Earth's crust, reaching the upper mantle (some 20–25 km beneath the surface). Meanwhile, its original explosion had produced searing global heat ejecting Gigatons of debris and dust into the atmosphere.

In addition, a globe-circling tsunami was then launched into the Gulf and beyond, drowning some 25% of the world's coasts and leaving a huge water-filled crater. Its crater water boiled into the atmosphere and altered regional oceanographic conditions. The resulting atmospheric contamination fueled a series of long-lasting climate disruptions. The totality of these short- and long-term changes resulted in the mass extinction of 60–80% of plant and animal life (including the dinosaurs) and likely changed the path of human evolution.

The Prologue describes Chicxulub's apocalyptic early morning fall to Earth 65.5 million years ago. Chapter 1 examines the nature of orbiting asteroids and their kinetic energies, followed by a discussion of historical asteroid impacts and the probability of future Earth–asteroid collisions.

---

[1] The object hit a much larger and deeper Gulf of Mexico at the location of what is now the coastal fishing village of Puerto Chicxulub—hence its name.

Chapter 2 describes the discovery of Chicxulub, beginning with the 1980 "Alvarez" analysis of the Cretaceous Paleocene (K-Pg) boundary, whose chemistry suggested that it was derived from a massive asteroid's impact on Earth. Years following the Alvarez expedition, the existence of a huge crater deep under the Gulf of Mexico was confirmed later as the asteroid impact site. A summary is given of the latest analyses of the Chicxulub data, also suggesting effects on the Earth's biosphere.

Chapter 3 paints a possible scenario of the Chicxulub impact sequence. Its explosion is described as it blasted through the atmosphere, the Gulf water, and deep into the Earth's crust. These explosive effects are compared to those of an underwater nuclear blast and include a model suggesting sequential formation of the impact crater. Similarities are drawn between the Chicxulub event and those of other asteroid impacts and volcanic eruptions. Finally, the energy of Chicxulub is compared with those of explosive geophysical phenomena and geological and astronomical events.

In Chapter 4, we discuss the tsunami created by Chicxulub and include a general "primer" on these types of waves and properties of refraction and run-up. Additional comments are made about historic earthquake and tsunami-generated floods. Two models of enormous tsunamis generated from different sources are then described. The first traces waves produced from a hypothetical caldera collapse in the Canary Islands. The second illustrates a future impact in the North Atlantic Ocean by an asteroid of 1.1 km diameter (one-tenth that of Chicxulub)—predicted to pass close to the Earth in 2085. Both models describe huge tsunamis hitting the shorelines that border the Atlantic Ocean. Finally, we draw a comparison between a laboratory-produced "Edgerton Effect" impact and the tsunami splash into the Gulf. We then track wave inundations around ancient Gulf coasts, the lowland areas of Central and North America, and then its rampage over the world's oceans.

Chapter 5 discusses long-term effects of the boiling Chicxulub crater water that fed atmospheric cloud as well as subsequently raising sea temperatures and altering oceanographic conditions in the Gulf and western North Atlantic Ocean. Meanwhile, the debris cloud blocking the sunlight shut down photosynthesis and produced frigid global temperatures and gray blizzards over the continents. As the atmosphere cleared, the remaining greenhouse gases dealt a final blow to Earth's biota, which suffered sweltering temperatures, perhaps for centuries.

Finally, a tally is made of the loss of biota from the mass extinction where approximately 60–80 % of all plants and animals were destroyed, ranging from the tiniest ocean foraminifera to land-lumbering dinosaurs. Mysteriously, some small mammals survived.

The Epilogue concludes our tale by an example of how the notable Cambrian Burgess Shale became isolated by pure chance, and subsequently precipitated animal extinctions, altering the tree of life. A question could be posed—how could an improbable single Chicxulub event alter by a similar process the entire future path of human evolution.

A Note from the Author (David Shonting)

The framework for this story was derived from research on tsunamis. As lead author and physical oceanographer, I was drawn to contemplating both the tsunami and the effects of the hot impact crater upon the ocean environment. Whetted by my interest in the study of waves and in the underlying sciences involved, I set about to revisit the Chicxulub story. Our tale touches upon an eclectic interplay of subjects including astronomy, geology, oceanography, chemistry, biosciences, and even world history. We have attempted to present the material not in a text format, but rather as a narrative, based on science for the general reader. Because this book relates to so many disciplines, tedious referencing has been minimized in order to help the reader engage with the flow of the writing. Alternatively, footnotes are used to expand on an idea or to provide added information.

A Note from the Coauthor (Cathy Ezrailson)

This Chicxulub tale starts out in fantasy and supposition and then initiates a journey that takes us through reality with its marvelous unfolding mysteries. Having had a rewarding career teaching young adults, I undertake this collaboration with my coauthor in order to help present a unique piece of Earth's history to the general public and to a new generation of science enthusiasts.

Naples, FL   David Shonting, Sc.D.
Vermillion, SD   Cathy Ezrailson, Ph.D.
January 2016

*This book is dedicated by DS to Bill Barrett, who nurtured
my interest in science and took me at age 8 to an MIT Open House,
and to Howard Wagner and Grace Farnum, my science and math teachers
at Laconia (NH) high school.
Also, this book is dedicated by CE to Ronald Stocker, my secondary
school physics teacher, who encouraged me to be tenacious; to Ray Lawson,
who taught me to write more succinctly; and Elfi Werzer, who instilled
in me the joy of learning.*

Acknowledgments

Much gratitude is given to our reviewers and copyeditors—those who have improved the reading and accuracy of this book through their meticulous edits: Tom Duncan, Alan Gibson, Susan Gapp, Lory LaPointe, Alan Massey, Carl Stendhal, and Mark Sweeney. Thank you to Alan Hildebrand and Steven Ward for use of their data and figures. All line drawings in this book were done by author David Shonting, unless otherwise indicated.

Contents

Abbreviations, Notation, and Metric Units

Dates quoted in the text are written in the form 01/Jan/1999 (meaning January 1, 1999). Since this book deals with physical quantities spanning great ranges of magnitudes, scientific notation (based on powers of ten) is used in order to avoid writing out multiple zeros or using terms such as trillion, quadrillion, etc. As an example, 500,000 and 0.00005 are written as 5.0×10^5 and 5.0×10^{-5}, respectively, where the exponent indicates the number of zeros to the right (+) or to the left (−) of the first digit. Since comparisons are frequently made of dimensions of objects or scales of phenomena, we employ the symbols > or < as greater than or lesser than and ~ as approximate values. Finally, Metric units are used throughout with some parenthetical notation, e.g., 17 km (11 mi).

Table of metric units used in the text

Quantity/Unit	Symbol	Equivalent/Comment
Length (l)		
Meter	(m)	Adult's stride
100 meters		Football field + end zones
Kilometer	(km)	10 of the above or 1000 m
Area (A)		
(Meters) ^{2}	$(m)^2$	Desktop
Volume (V)		
(Meters)^3	$(m)^3$	1000 L
		1000 kg (of H_2O)
		Metric ton (of H_2O)
(Kilometer)^3	$(km)^3$	~900 Rocks of Gibraltar
Mass (m) (Amount of matter in an object)		
Kilogram	(kg)	~2.2 lb
Metric Ton	1000 kg	
Gigaton	10^9 tonne	
Acceleration (a)		
Gravity (g)	m/s^2	~10 m/s^2, ~32 ft/s^2 (Earth)
Force (F) (Weight = mass × g)		
Newton	(N)	
Energy (E)		
Potential (Joule)	(PE)	Work to raise an apple ~1 meter
Kinetic (Joule)	(KE)	
Megaton (of TNT)	(Mt)	Million tons of TNT
Power (P)		
Watt (Joule/sec)	(W)	Raising an apple 1 meter per second

Prologue: The Arrival

Abstract: The Prologue describes Chicxulub's apocalyptic early morning fall to Earth 65.5 million years ago.

At the first sign of the object, a new star-like image appeared to move gradually across the night sky, its glow produced by reflected sunlight. Such was the beginning of the scenario that unfolded over subsequent days and led to that fateful event, when the giant asteroid crashed into the middle of the ancient Gulf of Mexico (Fig. P.1). Today, its impact site would be centered on what is the northern edge of the Yucatan Peninsula, a location near the little fishing village with the quaint Mayan name of Chicxulub (pronounced: chick-soo-loob).

Fig. P.1 Schematic of trajectory and landing site of the Chicxulub asteroid in the ancient Gulf of Mexico. The broken line suggests the coastline 65.5 million years ago. See Chapter 3 for details. Our vantage point "P" (pictured) is on the west tip of Cuba

Imagine that we are perched, within a protective time machine, high upon a 1000 m mountaintop on the ancient island that today is called Cuba (see Fig. P.1). There is a clear view to the western horizon where, on a calm patch of Gulf water some 400 km away, will soon become the Chicxulub object's "ground zero."

From our vantage point, the asteroid appears to grow in size and intensity each night, soon surpassing the brightness of the "evening star"—the planet Venus. Within 6 h of its early morning Earth impact, this disc will have increased, appearing as a smaller version of our full Moon. As its position changes relative to our vantage point and that of the sun, it begins to move into the Earth's shadow, briefly appearing as a brilliant crescent. Soon, the object becomes fully eclipsed, vanishing into a black void as if turned off by a galactic switch.

Just before the object's collision with Earth, it enters the thin upper atmosphere white ball streaked with yellows and reds. It continues to punch downward, as if a sudden quarter-sized Sun gone amuck (Fig. P.2). The night sky is lit up like a giant Roman candle.[1] Within seconds, the fiery object, with its blinding silver-white contrail, traverses its way through some 50 km (~30 mi) of atmosphere. This entire scenario is acted out as if in pantomime—in deathly silence.

As the object, still silent, dips below the horizon, its contrail starts to fade. Seconds later, the night sky bursts forth with brilliance greater than the brightest

---

[1] Such galactic fireworks, unfortunately [or perhaps fortunately], might only have been witnessed by creatures such as dinosaurs, some of which likely roamed the ancient Gulf coasts or on hillsides of western Cuba.

Fig. P.2 The early morning arrival of Chicxulub as might be observed from a mountaintop from our vantage point (P in Fig. P.1) on the west coast of Cuba

noon. Chicxulub had struck! This scene was perhaps similar to what humans later observed when the first nuclear ignition flash was triggered at the historic "Trinity Site" in Alamogordo, New Mexico.[2]

After several more seconds, a reborn white sphere began to rise brightly in the west—illuminating the horizon like the dawn of a giant sun. This fireball lofted skyward as its contents expanded outward with a brightness that fully ignited the early morning sky. By the time it nears the top of the atmosphere, its vertical motion slows to a standstill, as its intensity starts to fade. Expanding horizontally, the massive sphere flattens out, metamorphosing, within minutes, into a giant mushroom cloud whose sides sharply reflect the early morning sunlight.

Within ~2 min of the impact flash, this mute fireworks display is shattered by an intense rumble as the ground starts to vibrate wildly (like from the passage of

---

[2] The 1945 desert test of the first atomic bomb.

a thousand freight trains). Announcing the arrival of the earthquake-like pressure waves generated by the exploding asteroid, it then passed through the seawater, blasting through the sea bottom and finally embedding into the Earth's crust. The ground vibrations taper off within a few seconds and are followed by a short period of silence. Suddenly, these shaking waves arrive as low frequency rumbles as the ground oscillates tens of meters in amplitude. These waves had energies that could only have been generated by earthquakes far stronger than the largest ever recorded (such as the 9.5 Magnitude 1960 Chilean tremblors). The shaking lasted for several minutes, as rocks tumbled from steep cliffs producing giant landslides. Soon, all was again quiet.

Then, within 20–25 min, two crushing airborne pressure waves suddenly arrive.[3] The first is heralded by a mighty sonic "kaboom" from Chicxulub's atmospheric contrail. Even from our distant vantage point, this high decibel shock wave cuts through the atmosphere destroying all creatures' eardrums it encounters. A second or so later even more intense thundering and crackling airborne signals are felt from Chicxulub's Earth impact. These overpressures are sensed as hyper tornado-like 350 m/s (~750 mph) winds while adiabatic compression is causing air temperatures to soar hundreds of degrees. All around the vegetation, forests, and wildlife are simultaneously torn asunder and incinerated.

Meanwhile, up in the sky, an expanding mottled gray-black cloud builds into a great cumulonimbus mushroom cloud, as if generated by a monster volcano. Dark layers that resulted from the cloud's fine sooty ash and rock particles condensing from the thousands of cubic kilometers of Gulf water vaporize into a giant billowing plume. Thunderous rumblings also follow lightning flashes as intense storms form within and around this towering column.

Within minutes, as ominous clouds spread overhead, hot ejecta, accompanied by a rush of intensely heated air, began to rain down. This fusillade hailed the returned to Earth of asteroid and crater material that had been blown into the air by the impact. Most of these plunging missiles are incandescent containing huge rocks and boulders, some far exceeding the masses of large buildings. These became fresh new meteors, hitting the land and exploding as firebombs. Debris falling into the sea is visible as towering splashes far out from the beaches.

The mushroom cloud stirred by upper winds now spread, like the drawing of a giant gray curtain, and blot out much of the morning sunrise. Soon, in the gathering dark, new visitors from the sky arrive—the first raindrops. Rapidly intensifying, they soon coalesce into torrential downpours at rates not in the usual millimeters but

---

[3] Earthquake P and S waves travel in the crust at about 6.5 km/s and 3.7 km/s, respectively, while the speed of the sound in air is a mere 0.34 km/s, hence the earlier arrival of the earthquake waves. Such waves are further discussed in Chaps. 2 and 3.

Fig. P.3 The arrival of the towering Chicxulub tsunami waves over the fire-razed Cuban coast. The downpour of the muddy rains almost erases the giant luminous impact cloud to the West

in meters per hour. The dirty gray-black precipitation has a pasty consistency with an odor suggesting strong sulfur or acidic content. Soon, the countryside, already littered with shredded and charred trees, vegetation, and wildlife, become covered with poisonous muck—rapidly accumulating on sloping terrain, transforming into huge run-off mudslides of acidic slurries that race down into the sea.

As this scene of devastation around us seems complete, our attention is drawn to a deafening roar arriving from the open seas to the west. Out of the gray mists and driving rains, there looms a parade of gigantic waves—the Chicxulub tsunami (Fig. P.3). They seemed to defy the laws of hydrodynamics as their frothing white crests grow upward, approaching our eye level with wave heights several hundred meters above the sea. As the first waves crash ashore, with crests separated by wavelengths of over a kilometer, they seem to be tipping over onto us. These waves met and swallowed up the gray outwash run-off that now was 50–100 m deep.

Within minutes, the Gulf waters pour inland, drowning the foothills to our west. The crests now decrease to some 200–300 m and came to a halt at the base of our mountain vantage point. As the waves ebbed, they move in concert with the great outwash back into the sea, then again meet, and are swallowed up by the next even larger waves rushing shoreward. The entire inundation process repeats over and over until the oscillating waves exhaust their energy on the shores and finally die out. Meanwhile, the last of the giant tsunami waves still radiated out over the broad reaches of the ancient Gulf and subsequently into the Atlantic Ocean.

Thus ends the first act in the Chicxulub drama. From our perch, only the charred and gray landscapes remain—like fresh battlefields littered with destroyed life, sodden from pelting rains, and chilled from the darkened skies—all evidence of the massive asteroid's fatal collision with Mother Earth.

There is much more to this ancient Chicxulub saga and of its global consequences wreaked upon the Earth's environments. But, before we continue the story, it is instructive to back up a bit and examine the nature of asteroids in general—their physics, history, and the potential effects of their impacts upon land and ocean environments. Such background should assist the reader in comprehending the near-incomprehensible event—the great Chicxulub Impact.

Our story will be, as much as possible, based on available data, as well as conclusions from many scientists. However, it also inherently includes flights of imagination at times even approaching whimsey.

The original version of this book was revised. An erratum to this book can be found at DOI 10.1007/978-3-319-39487-9_6

1

The Orbiting Objects

Abstract This chapter defines and describes asteroids and their impacts with a description and comparison of their kinetic energies and impact crater formation. A description is made of the potential as well as kinetic energies involved. Accounts of two notable explosive impact events—Barringer and Tunguska are given. Also, NEO objects possibly (or probably) destined to have near catastrophic collisions with the Earth sometime in the future are described.

The Asteroids

An asteroid is an object in our solar system not large enough to be called a planet. Additional objects in our solar system include the planets, their satellites (moons) and comets all of which orbit our galactic shepherd and energy provider—the Sun.[1]

Our story is about the asteroid, Chicxulub a likely former member of an asteroid group in our solar system and its collision with Earth. In order to better comprehend Earth-asteroid impacts, we consider properties of these

---

[1] Comets, unlike asteroids, have larger elliptical solar orbits often extending beyond the solar system. These "dirty snowballs" can have a nucleus of rocks, frozen gases, and ice, held together by the comet's weak gravity. When near the Sun this material heats up and the Sun's radiation pressure pushes the gases away from the nucleus forming the tail or coma, thousands or even millions of kilometers long, where its reflected sunlight, glows in the night sky. Comets, which can move up to twice as fast as asteroids, make up only a few percent of orbiting objects.

The original version of this chapter was revised. The erratum to this chapter is available at:
DOI 10.1007/978-3-319-39487-9_6

© Springer International Publishing Switzerland 2017
D. Shonting, C. Ezrailson, *Chicxulub: The Impact and Tsunami*,
Springer Praxis Books, DOI 10.1007/978-3-319-39487-9_1

sometime intruders from space. These orbiters range from sand-sized particles of space dust to the enormous object Ceres, 945 km (590 mi) in diameter.[2] Objects less than ~50 m (150 ft) in diameter, are generally referred to as meteors. If they happen to approach the Earth and make it through the atmosphere to the ground, they are then called meteorites. About 15 % of recovered meteorites are metallic in composition, and about 85 % are the stony type referred to as chondrites, usually containing small grains of fused silicate minerals and metallic oxides called chondrules. Asteroids are 4–5 billion years old and are considered the building blocks of our solar system.

Most of the orbiting objects within our Solar System reside in the Main Asteroid Belt (see Fig. 1.1A), lying between the orbits of Mars, and Jupiter. But, because they are outside the Earth's orbit they are considered no threat to us. This is fortunate because this group contains many large asteroids (Table 1.1). Were any of these asteroids, say >15–25 km, to collide with Earth, the result would be catastrophic, due to the incredible energy that would be released at impact. Such an event would likely kill all life forms inhabiting the Earth.

Two additional groups of asteroids (also shown in Fig. 1.1), the members of whom could intercept the Earth's path about the Sun, have either long elliptical orbits (B) or smaller, more circular paths (C). The majority of these objects are less than ~100 m in diameter, but many range up to several hundred meters with a few even reaching over a kilometer in diameter. The larger of the asteroids in groups (B) and (C) are categorized with the unsettling title of "Near-Earth Objects" (NEOs).

With the aid of newer, more powerful telescopes, the list of solar orbiting objects being tracked—especially the NEOs—is (unfortunately) steadily increasing. Over 1000 NEO's have been identified. However, statistics suggest that there are several thousand more of the 0.5–1 km or larger objects still out there, as yet undiscovered.

Asteroids were viewed with only Earth-bound telescopes until in 1991 when the oversized asteroid Gaspra was actually visited by the Galileo spacecraft. Since that time, Galileo has collected many high-resolution photographs (Fig. 1.2).[3] Gaspra's shape resembles a giant Idaho potato, roughly 18 km long and 10 km wide. A comparison with the Empire State Building in New York City allows us to contemplate what would happen if Gaspra were to collide

---

[2] Although some asteroids are not particularly spherical, for simplicity we'll refer to diameters as a useful scale.

[3] The Galileo (NASA spacecraft) was a most successful explorer of Jupiter, its moons, with flybys to asteroids Gaspra and Ida (with its tiny moon). After far-flung searches, Galileo, so as not to bio-contaminate any of Jupiter's moons, sadly was crashed into the red planet.

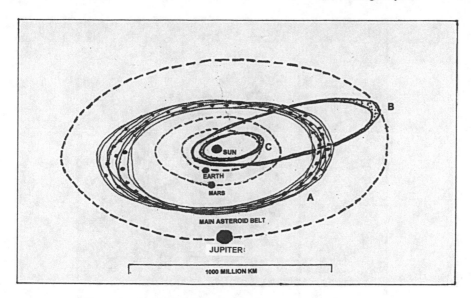

Fig. 1.1 The three asteroid belts around the Sun. *A* is the near circular Main Asteroid Belt whose orbits do not intercept that of the Earth. The more elliptical belts *B* and *C* contain objects (including NEO's) whose orbits can cross the Earth's orbit

Table 1.1 The larger members of the main asteroid belt

Asteroid	Mean diameter (km)	Mass (tonnes × 10¹²)
Ceres	945	870,000
Pallas	520	318,000
Vesta	520	300,000
Juno	240	20,000
Eugenia	226	6100
Chiron	180	4000
Siwa	103	1500
Mathilde	50	103
Ida	40	100
Eros	25	7
Gaspra	14	10
For reference		
(Earth)	(12,756)	(6.0×10^{21})
(Moon)	(3746)	(7.0×10^{19})

with it. But, as we have already noted, this and most of the other outsized objects reside in the Main Asteroid Belt, safely out of range of the Earth's orbit.

Thousands of meteoric objects enter the earth's atmosphere daily; most of these ranging from pebble—to boulder size. A most remarkable property of these objects is their incredible speeds, which commonly range from ~15 to

Fig. 1.2 The giant asteroid Gaspra photographed in Oct/91 from the visiting spacecraft Galileo (NASA photo). *Insert*: The New York Empire State building roughly to scale

50 km/s (~34,000–112,000 mi/hr). When a meteor enters the atmosphere at such speeds, air friction heats up their leading surfaces to incandescence while slowing them down—not unlike the Space Shuttle whose re-entries taught us the necessity of using heat shielding ceramic materials.

Meteors with masses larger than a few kilograms appear on clear nights as glowing white streaks and are called "bolides" or "shooting stars." There is, however, a more ominous aspect of orbiting asteroids—the possibility one of significant size colliding with Earth.

Asteroid Impacts: It's All About Kinetic Energy

Most asteroids, especially the stony variety with diameters less than ~50 m (165 ft), simply explode and burn up in the Earth's atmosphere, and whose remnants settle out as fine dust. Asteroids larger than 50–75 m (which, fortunately, are far fewer in number), because of their great momentum at atmospheric entry, impact the Earth—where they disintegrate into fragments. Since their initial formation, the Earth, Moon, and the other planets (and their moons) have been pummeled by these objects. Consequently, myriads of impact craters have pockmarked celestial objects with a solid surface.[4] On Earth, however, only a mere ~200 craters have been identified, belying the true frequency of impact events, since over the millennia weathering and natural geological processes have removed or covered most land craters.

Over the past 100 million years or so, 70–80 % of the Earth's surface has been covered by ocean, ensuring that most impacts occurred into water. In these cases, the energy would be largely absorbed in the creation of an explosion and its immense splash. Such splashes, especially in deep water, would mutate into huge tsunamis. In that case, little energy would make it to the sea bottom, explaining the dearth of undersea craters.

To convincingly describe the scenario of a medium-sized asteroid hitting the Earth, whose dimensions, speeds, and energies far exceed the range of human experience—is indeed a bewildering challenge. To wit, we will try to visualize a Gibraltar–sized boulder racing across the sky at a speed of say 20 km/s (~45,000 mi/hr).[5] And then, craft our scenario associated with the object's crashing into Earth—such would stretch the imagination, of even a science fiction writer.

What actually happens in such collisions is all about the release of Kinetic Energy—the energy of motion. Kinetic Energy (KE) and its complementary quantity—Potential Energy (PE) can be illustrated in the following way: assume we have a stone of mass M (kg) resting atop of a cliff of height h (m) above a pond (see Fig. 1.3). This stone presses down on the cliff with a force equal to its weight—defined as its mass M multiplied by the acceleration due to gravity g.

This would be written as M × g or just Mg. With reference to the surface of the pond (h meters below) the stone is said to have a potential energy (PE) that is equal to its weight Mg multiplied by its height h above the pond, or

$$PE = (\text{weight}) \times h = Mgh \qquad (1.1)$$

Think of the PE as stored energy equal to the amount of work you would perform if you lifted the stone vertically from the pond level (at h = 0) to the top of the cliff at height h.[6]

---

[4] Our outer giant planets, like Saturn and Jupiter, have gaseous or semi-liquid surfaces and hence do not permanently display impacts.

[5] Compare this speed with those of the orbiting space shuttle or a high-powered rifle bullet moving at 6–8 km/s.

[6] I still recall my physics professor, Jonathan Karas, relating that if you carried the stone back down to the pond level, the net amount of work or energy the stone gained for the up and down trip—would be zero, no matter how exhausted you were.

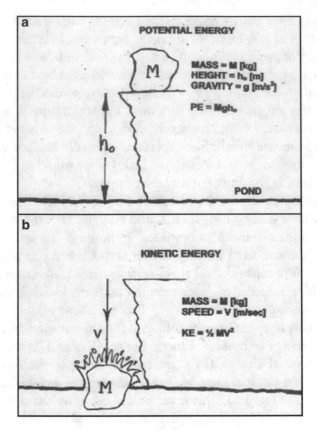

Fig. 1.3 (a) A stone resting on a cliff over a pond whose surface is h_0 meters below and has the Potential Energy PE. (b) The stone hitting the water having gained Kinetic Energy KE

Now, if we push the rock over the cliff, under the influence of gravity, it begins to accelerate as it falls. And since the value of g is ~10 m/s/s (written as m/s²), the stone will increase its speed by ~10 m/s (32 ft/s) for each second it falls.[7] At the instant the stone hits the pond's surface, it has fallen through the distance h and it is moving at a speed V. During its fall, it has accumulated an amount of KE that, at any instant, is defined as proportional to its mass M times the square of its speed or V² (V multiplied by itself) or

---

[7] An acceleration of ~10 m/s² means that a falling object (in a vacuum) increases its speed by ~10 m/s every second. In air, the object soon attains terminal velocity as the frictional force approaches that of gravitational force.

$$KE = \frac{1}{2}M(V)^2 \tag{1.2}$$

where ½ is proportionality constant. Note that the term $(V)^2$ means that if you double an object's speed you quadruple its KE—a non-linear effect.

Recapping: resting on the cliff at height (h) the stone possessed maximum PE but zero KE (since its initial velocity $V = 0$ in Eq. (1.2). Then, when the falling stone just impacts the water it has maximum speed, i.e., maximum KE. However at this instant the stone has zero PE (since now its height $h = 0$ in Eq. (1.1). Thus, in falling from the cliff, the stone's entire PE has changed into KE.[8]

When a rock hits the water or ground, it is the sudden deceleration (negative acceleration) that transforms the KE into other forms of energy. For a stone, it's simply a splash or a noisy clatter! But, what about a stone, moving at the unearthly speeds of the asteroids? To explore this question let's assume a falling spherical stone two meters (~6 ft) in diameter with a density of 3000 kg/m³ (three times as dense as water—about that of common Earth rocks). It would hit the pond at a speed of 45 m/s (~100 mi/hr). The mass M of this stone is its density times its volume (4/3 π R³), where the radius R is 1.0 m. From Eq. (1.2) the KE is ~12,000,000 J (written as 1.2×10^7 J).[9]

Suppose that our stone is instead a speeding meteor, now its Earth impact would be an incredibly bigger event. Assume it is moving not at 45 m/s but at a typical asteroid speed of 20 km/s, its KE would now be 2.4×10^{12} J—over 200,000 times greater than that of the stone falling to the pond, or equivalent to ~500 tons of TNT. Some splash!

On Kinetic Impacts

The reader may be uncomfortable with the fact that there can be such large explosions caused by a simple impact of a rock with land or water—when there are no apparent chemical or electrical explosive materials present. So, let us consider the case where an object falls into water. First, there is an enormous difference between the energy-releasing mechanisms in the impacts of the falling stone compared to that of a racing meteor. A relatively slow falling stone transfers most of its KE into the turbulent splash in which the rock's volume merely displaces and throws an equivalent volume of water into the air.

---

[8] One can think of potential energy as like having money in the bank whereas kinetic energy is going shopping with a debit card.

[9] The Joule is the metric unit of energy and is equivalent to the work required to raise an average-sized apple 1 m.

When a meteor impacts the water with incredibly greater speeds, much of the water, because of its inertia and its viscosity, can't get out of the way fast enough. So, in this case, the molecules and atoms of the object as well as of the target material have no choice but to squeeze together. This collapse forms liquid-like plasma, which generates instant temperatures of thousands of degrees. The atomic and molecular forces that act like tiny springs now resist the compression of molecules and atoms and these super powerful coiled springs having built up huge pressures finally blow outward with an explosive force. This effect is the same for an asteroid whether it impacts land or water. The result is called a "kinetic explosion" which is "fueled" by pure kinetic energy, that manifests itself by producing heat and pressures far greater than in a chemical explosion, or even a volcanic eruption and more like the results akin to a nuclear bomb—except without the radioactivity.[10] The size of such man-made explosions (chemical or nuclear), are limited by the amount of explosive or nuclear material available, while a volcano is triggered by built up pressures within the Earth. Even these kinetic energies can be dwarfed, however, by an asteroid's impact whose kinetic energies are defined by its mass and speed, and can vary by immense ranges.[11]

To better comprehend the magnitudes of the energies of events such as asteroid impacts, volcanic eruptions or even nuclear explosions, scientists define these values in terms of millions of tons (megatons or Mt) of exploding TNT where one Mt is equivalent to 4.2×10^{15} J. The atomic bomb dropped on Hiroshima produced ~0.015 Mt whereas the largest hydrogen bomb test (the Russian "Tsar Bomba") generated ~50 Mt.

Figure 1.4 illustrates the huge range of magnitudes of asteroid energies. It compares the KE's, with the small falling stone and the speeding asteroid with two of the larger Earth orbit-crossing asteroids—Apophis, and 1950DA. The ordinate axis gives the order of magnitude (in powers of ten) of the KE values in Joules. The estimated speeds of the two asteroids were obtained from radar and optical tracking. Both are predicted to come extremely close to Earth within the next 75 years and are discussed at the end of this chapter. Compared to the small meteor the KE's of the two asteroids are from one million to a hundred million times as great. Moving from hypothetical asteroid impacts we next examine two actual object-Earth collision events.

---

[10] Chemical bomb energy is produced by the rupturing of chemical bonds and instantly creates a huge volume of gases, producing high temperatures and pressures. Similarly, but more energetic, is an atomic bomb produced by the release of nuclear bonds, whose heat and radiation vaporize the bomb materials creating much higher temperature and pressures.

[11] Be reminded that the mass of an asteroid (thus its kinetic energy) varies with the CUBE (r^3) of its radius.

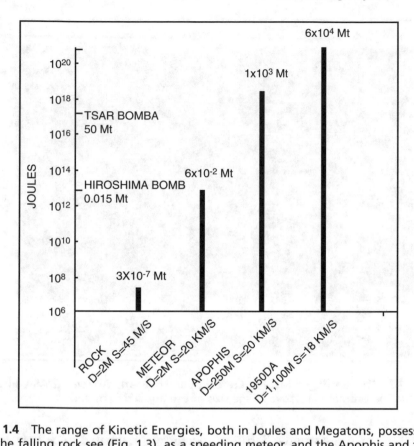

Fig. 1.4 The range of Kinetic Energies, both in Joules and Megatons, possessed by the falling rock see (Fig. 1.3), as a speeding meteor, and the Apophis and the 1950DA asteroids. Also shown for reference is the Hiroshima Atomic bomb and the largest Russian Hydrogen bomb

The Barringer Impact Crater

The well-preserved Barringer Meteor Crater (Pilkington and Grieve 1992) near Flagstaff, AZ (Fig. 1.5) demonstrates the starkest evidence of the kinetic energy released in a meteor crash to Earth.[12] Occurring some 50,000 years ago, the impact produced a huge cavity ~1200 m (3600 ft) in diameter and ~200 m (600 ft) deep having a volume of ~0.2 cubic kilometers (about 80 Louisiana Superdomes). Excavations around the crater suggest that the impacting object was a metallic iron-nickel meteor, which had exploded into

---

[12] Named for Daniel Barringer, a mining engineer who worked to prove that the crater was not volcanic but the result of an asteroid impact. His case was verified by geologist Eugene Shoemaker from the University of Arizona who showed that fused crater rocks (called tektites) were similar to those found in Nevada nuclear test sites, evidence of the occurrence of intense heat and pressures.

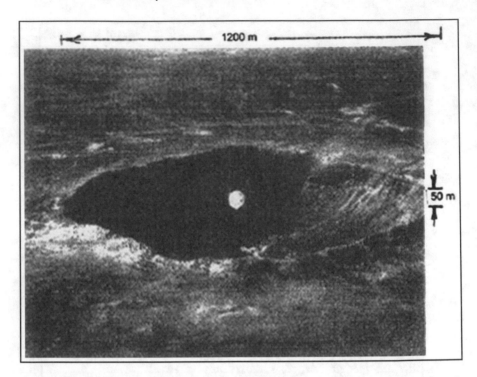

Fig. 1.5 The Barringer Meteor Crater, near Flagstaff, Arizona (NASA Photo). Insert: The estimated relative image size of the impacting meteor

small fragments. Since the Barringer meteor was estimated to be only about 50 m (150 ft) in diameter, it seemed incomprehensible that this relatively small object's explosive effects could so grossly distort the desert landscape. The meteor sketched in the photo (Fig. 1.5) is roughly to scale, its volume was clearly a tiny fraction of the crater material blown skyward. This is a vivid testament to the validity of a typical physics 101 lesson on the release of kinetic energy.[13]

In order to get a better feel for the Barringer's kinetic energy, let's assume its speed V was 20 km/s with a spherical volume of ~62,300 m³—roughly that of a fifteen story office building. Further, assume the object's density was ~9000 kg/m³ typical of a metallic object. This gives a mass M of 560,800 tons—more than five Nimitz Class aircraft carriers. Plugging the speed V, and mass M, into Eq. (1.2) yields a KE of ~1.2×10^{17} J. This is the same order of magnitude of the energy as 1930 Hiroshima bombs or 29 Mt (see Table 3.2 and discussion in Chap. 3).

---

[13] This is equivalent to a baseball-sized rock (7.5 cm or 3″ diameter) blasting a crater in concrete 1.8 m (6 ft) in diameter.

Within milliseconds of impact, the Barringer's KE was released to the hard desert floor. Its compressed molecules then rebounded, forming the kinetic energy explosion (discussed above), completely shattering the asteroid to bits. This would have produced a shock wave of searing heat far exceeding tornado-speeds with winds of ~1000 km/h (~625 mi/hr) extending tens or hundreds of kilometers outward, and decimating any organic matter in its path. The shock wave was followed by millions of tons of hot ejecta thrown high into the air.

Such an explosive force, associated with such a relatively small asteroid, suggests that if, instead, a similar object blasted into the open ocean, a cavernous water crater would be formed, the collapse of which would then create a monstrous tsunami hundreds of meters high (see Chap. 4 for further discussion).

The Tunguska Explosion

A more recent and perhaps larger object impact occurred near the Tunguska River in desolate western Siberia in 1908. This was one of the first major objects of note to hit the Earth during recorded human history. However, because of its remote location, this event remains pretty much a mystery. Prof. Leonid Kulik, the Russian mineralogist who led one of the many site expeditions, suggested that the notable absence of a crater and the paucity of fragments recovered, indicates that the object may have been made up of an aggregate of rocks and ice i.e. may have been a comet. This object, with an estimated diameter of 60–80 m (180–240 ft), apparently exploded high in the air, creating a fireball whose heat radiation incinerated vast areas of vegetation and forestation. Estimates of the KE (dissipated from the explosion effects on the land) ranged from 8 to 10 Mt or some 500–600 Hiroshima bombs. The pressure wave, which flattened some 80 million trees over an area ~50 km in diameter (see Fig. 1.6) was recorded on barographs on all continents and produced a thunderous echo in London, half way around the Earth.[14]

It is evident that, although the Barringer object was denser and potentially more powerful than Tunguska, its energy was more concentrated—directly into the impact site. In contrast, the Tunguska's in-air blast spread energy and

---

[14] The recording barometer registers continuous changes in atmospheric pressure.

Fig. 1.6 Results of Tunguska object air—explosion in 1908, which scorched and flattened over 2000 km² of Siberian Forest

destruction over a much larger area.[15] That Tunguska happened in such a remote Earth location was a fortunate luck of the draw. Had the impact happened over London, this bolide would have been seen all over Britain, brighter than the Sun followed by dark contrail. The impact explosion would have destroyed the city killing millions with damage extending out hundreds of kilometers throughout the British Isles—creating the worst natural catastrophe in recorded history.

On the Probabilities of Asteroid-Earth Collisions

We humans are of course concerned about the probability of an asteroid (especially a large one) hitting Earth. To explore this question, scientists have tabulated larger observed asteroids (including the NEOs) in the Earth-crossing asteroid belts B and C (see Fig. 1.1). Noting their size distributions compared to actual Earth—impact events their probable frequency of occurrences have been estimated and are summarized in Fig. 1.7.

---

[15] Since no obvious crater was ever found, the cause of the event was subject to wide (and indeed wild) speculations, from that of a black hole passing through the Earth, to the arrival of some anti-matter, even to that of crash of a wayward E.T.-like spaceship.

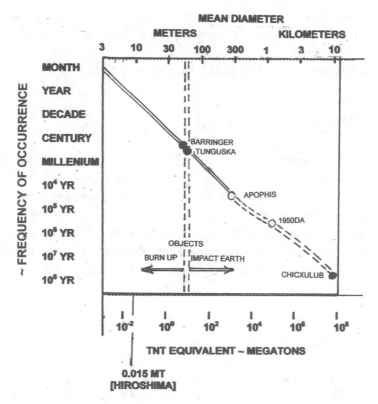

Fig. 1.7 Estimated frequencies of impacts for given sizes and kinetic energies of the asteroids (Re-drawn from Chapman and Morrison, 1994). The graph is very rough because of the uncertainty of speeds and rarity of events (especially in the range of the *broken line*). It is suggested that frequencies of occurrence and energies for objects ≳100 m diameter (or ≳100 Mt) may err by a factor of three

Shown also are rough estimates of the impact kinetic energies calculated assuming a 20 km/s impact speed and an object mean density of light rock, ~3000 kg/m^3. The Fig. 1.7 includes three known asteroid impacts (black circles): the Barringer and Tunguska events and at the far end of the scale is the giant "Chicxulub" that had miniscule probability of happening once in a hundred or so million years. The reader is warned that this probability of impact should not be interpreted as a prediction. It only suggests that if we have records of several impacts of a certain size asteroid, then we can estimate the average time between impacts. However, if we have only a single event, Chicxulub, in 65.5 million years, we have insufficient data from which to calculate an average impact interval. If we knew how many Chicxulub—sized objects were out there, and knew their orbits then we could estimate their mean frequency of occurrence. But, of course we don't and can't.

Clearly, the larger the asteroid, the less frequent is the probability of their impacts. This is consistent with the fact that the population of observed objects is heavily skewed toward smaller sizes, which suggests why the Tunguska Impact was the largest of its size in recorded history.

Other Objects of Concern

Our probability graph shows two "real" NEO asteroids whose forecasted future proximities to Earth are of concern. The first object "Apophis", as discussed above and sketched in Fig. 1.8, is shown next to the Empire State Building. Under the watchful eyes of several telescopes, this 260 m diameter—66 billion ton rock swings around its elliptical solar orbit in the "C" group of asteroids. The problem with Apophis is that its orbital plane is similar to Earth's and it has period of 323 days. As a result, it crosses near the Earth's orbit roughly twice per year. Even worse—astronomers predict it to pass within only ~31,000 km of Earth on 13/Apr/2029 at 04:36 Greenwich Mean Time with a speed of ~30 km/s (67,500 mi/hr). With this speed and a density of ~3000 kg/m^3 the object's KE is about 3000 times the KE of the Barringer asteroid.

Apophis would be seen as a bright star (reflecting light from the Sun) moving through the constellation Cancer. This could be the first asteroid to be visible (outside the Earth's atmosphere) with the naked eye. Normally such a flyby would be a rarity, not reoccurring again for millennia. There may be a glitch here, however! Astronomers fear that with Apophis' passing so near to Earth, that Earth's gravity might tweak Apophis' orbit just enough to cause it to pass even closer, or even collide, when it returns 7 years later—and coincidentally on 13/Apr/2036. Newer observations of Apophis suggest its probability of Earth impact is only ~1 in 45,000 (0.002 %). Even with this low value, astronomers have urged NASA to "fine tune" Apophis' proximity prediction, by perhaps planting a transponder on it for close tracking. Or, even (as Hollywood might portray it) to prepare to alter its orbit by nudging it with explosives from a spacecraft—now considered a remedy, if given enough time to react.[16]

Projections of Apophis' impact have kept the computer modelers busy. One prediction is for it to hit Earth within a 50 km wide swath beginning in eastern Russia and ending over the Caribbean. The best money however, has

---

[16] At the time of this writing the European Space Agency has just recently landed the Philae probe on the comet 67P/Churyumov-Gerasimenko. This remarkable feat suggests altering an asteroid's orbit is feasible.

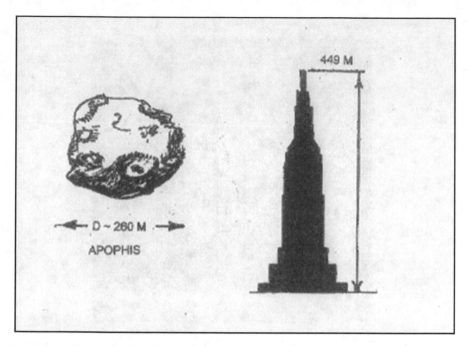

449 M

D ~ 260 M

APOPHIS

Fig. 1.8 Sketch of Apophis forecasted to come close to earth in 2029. *Insert*: The Empire State building to scale

it slamming into the central Pacific, where with water depths over 3000 m, it would blast a water crater 7 km wide and 2700 m deep. Its collapsing walls would send a tsunami racing over the entire Pacific. And within hours the coasts of the Pacific Rim countries would be pounded with wave heights of 15–20 m (45–60 ft). Stay tuned for this one!

Another perhaps more serious threat to Earth, but in the more distant future, is the NEO "1950 DA," a 1.1 km (1100 m) giant, discovered in 1950.[17] It has been tracked both with optical telescopes and radar, the latter presenting a quite spherical but rather spooky image (see Fig. 1.9). It is predicted to approach uncomfortably close to Earth (within a lunar distance ~384,000 km) in 2880. Worse, in 2002 Jon Georgini from the California Jet Propulsion laboratory and several others, published in the journal *Science*, suggested a likelihood of actual Earth-impact probability of a discomforting 0.3%.

For this giant object, much larger than Apophis, an Earth-impact would cause a much more catastrophic disruption wherever it landed. For an impact

---

[17] NEO's are listed by the year discovered followed by the observer's initials.

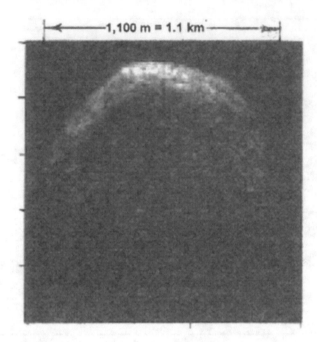

Fig. 1.9 A spooky radar image of asteroid 1950 DA forecasted to approach Earth in 2089 (NASA Photo)

in one of the world's oceans, computer modeling predicts tsunami waves hitting the continental coasts exceeding ~100–150 m high.[18] A model of an ocean impact of 1950DA is given in Chap. 4. To place a perspective upon the sizes of the two "proximity" objects along with Tunguska and Barringer, Fig. 1.10 suggests the comparison of each object with Chicxulub.

Then, there is the problem of the "party crashers". Aside from the tracked NEOs, astronomers are still perplexed with rogue asteroids that suddenly arrive in the Earth's backyard unannounced. For example, a surprise flyby occurred on 04/Mar/2009 when a meteor, 30 m in diameter, was detected by NASA radars and according to the evening news, came within 76,800 km (48,000 mi) of Earth—about a quarter of the distance to the Moon.

Our nominee for the scariest surprise event in recent years occurred on 23/Mar/1989 when NASA observed an asteroid ~400 m in diameter passing within 640,000 km (400,000 miles) of Earth. If that was not bad enough, the object, weighing about 300 billion (3.0×10^{11}) tons and speeding at 20 km/s

---

[18] Dr. Steven Ward from the University of California/Santa Cruz has prepared a series of fascinating computer model on-line displays of the creation of tsunamis by asteroid impacts, Earthquakes, volcanic eruptions, and landslides. See http://es.ucsc.edu/~ward/.

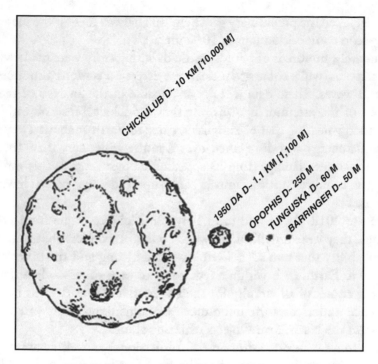

Fig. 1.10 Approximate size comparisons of asteroids Apophis, 1950DA, Barringer, and Tunguska with the giant Chicxulub

(~45,000 mi/hr), had crossed in the vicinity of the Earth's orbit undetected—just 6 h earlier! Cheers!

And finally, on 15/Feb/2013, a most remarkable extra-terrestrial coincidence occurred. An NEO asteroid designated as 2012 DA14 which had been tracked for a year, raced over Sumatra, Indonesia at some 28,000 km, passing inside the orbits of weather/communications satellites. The ~40 m object weighed about 100,000 tons and was the largest object observed to pass so close to Earth.

But wait—there is more! Most incredibly about 12 h earlier, citizens of Chelyabinsk, Byelorussia looking up to the morning sky observed a bright white trail streaking across the horizon (Chang 2013). About 3 min later they were buffeted by an intense shock wave that shattered windows, causing over a thousand injuries. Astronomers estimate the object was some 20 m diameter and entered the Earth's atmosphere at ~18 km/s. Its whitish trailing tail commenced at about 40 km altitude. And at about 15–20 km, it appeared to partially explode with an energy estimated at ~0.5 Mt (37 Hiroshima bombs). Downrange from the explosion meteorite fragments were found in the snow—mostly pebble-sized rocks however one larger piece (>100 kg) landed in a frozen lake. The preliminary geochemical analyses suggest the meteor was of stony-iron composition (about 10 % metal).[19]

Fortunately, hundreds of photos of the dark trajectory were made with cell phone cameras, while some 45 nuclear test detection stations also monitored the sound waves. These data will prove valuable in the analysis of the exact trajectory of the asteroid. It turns out that the Chalyabinsk object, having kinetic energy nearing that of Tunguska, came to Earth at about a low (~20°) angle so its energy was distributed over a much wider area than Tunguska, causing much less damage from its supersonic boom. Again, this was fortunate since a more vertical impact and upon a city, could have been disastrous.

Since the 2012 DA14 and the Chalyabinsk objects came from different directions, they were unrelated. However, one still wonders about the minuscule probability that two such asteroids—one hitting and the other coming very close to Earth, each within a few hours of each other—when there was such a big chunk of all eternity for their appearance. Clearly, this incredible coincidence underscores the importance of humankind paying much more attention to the Near Earth Objects whizzing about.

If it is some comfort, astronomers (professional as well as amateur) are tracking the known, and scanning the heavens for as yet undiscovered NEOs. Moreover, in 1998 the US Congress mandated NASA to catalog objects over a km in diameter. Then in 2005 Congress upped the anti—asking NASA to identify at least 90 % of the NEO's larger than 140 m by 2020 (a minimum size able to severely damage a population center). The United States, the European Union, and other nations as well as groups such as the Planetary Society have meanwhile created alliances (e.g., Spaceguard ll) in order to promote NEO tracking.

References

99942 Apophis (2007) Wikipedia. http://en.wikipedia.org/wiki/99942_Apophis. Accessed 28 Aug 2007

Chang K (2013) The week (story of Chalyabinsk asteroid). New York Times 19:A12

Chapman CR, Morrison D (1994) Impacts on the Earth by asteroids and comets: assessing the hazard. Nature 367(6458):33–40

Georgini J (2002) Asteroid 1950 DA's encounter with earth in 2880: physical limits of collision probability prediction. Science 296:132–136

---

[19] This object had been undetected, probably because it was coming from a direction close to the sun, making optical sighting difficult.

Napier W, Asher D (2009) The Tunguska impact and beyond. Astron Geophys 50:18–26

Near Earth Object Program (2015) NASA. http://neo.jpl.nasa.gov/risks/. Accessed 15 Aug 2012

Pilkington M, Grieve R (1992) The geophysical signature of terrestrial impact craters. Rev Geophys 30:161–181

Ward S (2009) A tsunami ball approach to storm surge and inundation: application to hurricane Katrina. Int J Geophys 2009, 324707. doi:10.1155/2009/3247

2

The Tale of Chicxulub

Abstract After reviewing the character of asteroids, we now turn to the story of the discovery of the remains of the great Chicxulub object that fell to Earth, some 65.5 million years ago. Piecing together the data, scientists discovered worldwide evidence in the K-Pg Boundary portraying the mass extinction of 60–80 % of the life on Earth. Some 20 years later, further findings exposed the crater along with evidence of the effects of the impact from outcrops in ancient Gulf of Mexico and deep bore holes around the world. Also, described is the controversy in which claims that Chicxulub's extinctions were caused partly or only by the eruptions in the Indian Deccan plateaus.

The Alvarez Discovery: The Origin of the K-Pg Boundary

This part of our story begins in 1980 with a publication by scientists from the University of California, Berkeley, and the Lawrence Berkeley Laboratory (Alvarez et al. 1980). This group which, included the retired Nobel Laureate, physicist Louis Alvarez and his geochemist son, were examining ancient Earth rocks for evidence of past life extinction events. Their focus centered on exposed limestone formation in the K-Pg Boundary from an ancient deep-sea bed found in an eroded cliff near Gubbio, a medieval town in Northern Umbria, Italy.

The original version of this chapter was revised. The erratum to this chapter is available at:
DOI 10.1007/978-3-319-39487-9_6

© Springer International Publishing Switzerland 2017
D. Shonting, C. Ezrailson, *Chicxulub: The Impact and Tsunami*,
Springer Praxis Books, DOI 10.1007/978-3-319-39487-9_2

The limestone was formed in part from tiny (~0.1–2 mm) fossilized calcium carbonate shells of benthic foraminifera (forams). These organisms have evolved in the world's oceans over eons into thousands of species, many of which had developed unique shell forms during different geologic ages. Their shells and bony structures settled onto the ocean floor and over millions of years became compressed and fossilized, consolidating into much of the Earth's limestone deposits.[1] By vertically sampling carbonate sediments in the deep ocean as well as limestone from ancient uplifted sea floors, paleontologists have used forams to date rock layers and to reveal changes in marine environments (e.g., temperature and climate) occurring over millions of years.

In multiple locations, the Gubbio rock strata exhibited two rock layers each with distinctly different foram populations. The deeper (i.e., the older) layer formed from a dense concentration of foram shells, displaying a large diversity of species. The upper (younger) strata of rock contained a less dense fossil concentration with different varieties of forams. Between the layers lay a thin intermediate clay boundary that would become one of the most significant paleontological markers in geologic history.

The clay interface between these two foram layers ~1–2 cm thick, contained no carbonate fossils. Potassium-Argon radioactive dating placed this layer's age at 65.5 million years.[2] This fell precisely between the late Cretaceous and the beginning of the Paleocene Period known as the "K-Pg boundary."[3] This boundary had been established in the 1970s as an indicator (in geologic time) of the most recent and extensive of the great biological extinctions occurring within the last 500 million years.

For a perspective of the time of the formation of the K-Pg boundary Fig. 2.1 displays the traditional Geologic Column (drawn from Leets and Judson 1958). Shown are five major mass extinction periods. The K-Pg boundary divides the late Cretaceous from the Paleocene period when the fossil records indicate the sudden disappearance of the dinosaurs.[4]

In order to explain the K-Pg extinction many hypotheses have been proposed. Scientists discussed this issue at several international meetings

---

[1] This type of limestone was used to build the pyramids of Egypt.

[2] Potassium-Argon dating is based on the measurement of the product of radioactive decay of an isotope of potassium (K) into argon (Ar). Potassium is commonly found in micas and clays. The decay product Ar-40 (a gas) escapes in liquid (molten) rock, but starts to accumulate when the rock solidifies (recrystallizes). The time since solidification is calculated by measuring the ratio of Ar-40 accumulated to the amount of K-40 remaining. The long half-life of K-40 allows calculation of the absolute ages of samples in the millions of years.

[3] Paleontologists tend to keep changing their time scale names. What was referred to as the Cretaceous-Tertiary (K-T, noting Cretaceous in German is spelled with a K) boundary now is called the Cretaceous—Paleogene (or Paleocene) or K-Pg boundary.

[4] We note that the period of Homo sapiens existence has occurred over only ~3 million years, far less than then the ~200 million years rein of the dinosaurs.)

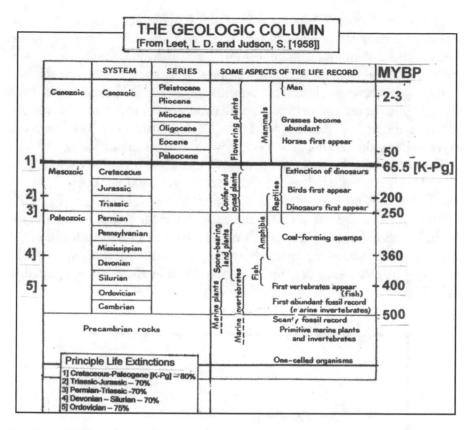

Fig. 2.1 The geologic column (redrawn from Leets and Judson 1958), showing the late Cretaceous—Paleocene (K-Pg) boundary (*dark line*) marking the life extinction. The *insert* shows the important life extinction events with percent of species destruction

including in Ottawa (1976) and in Copenhagen (1979), but arrived at little agreement as to its causes. Theories floated about included: (1) traumatic changes in oceanographic, atmospheric, or climate conditions which had been driven by continental-sized volcanic eruptions such as the Indian "Deccan" events discussed at the end of the chapter, (2) gross magnetic field reversals and polar shifts; and even more bizarre events, such as (3) those caused by a nearby super nova explosion, or (4) a possible flood from a giant underground Arctic lake.[5] The problem with these hypotheses was that all lacked direct unambiguous physical evidence.

---

[5] Supernovas are huge exploding stars. Those which have aged collapsing under their own weight, generate immense thermal energy from gravity, reaching ~45–50 billion degrees C. then exploding, flinging mass into space. Such material appears to have different isotopic ratios than that of our solar system.

The Alvarez geochemical analyses of the Gubbio K-Pg layer were most revealing: it contained unusually high concentrations of the platinum group elements, i.e., platinum, iridium, osmium, and rhodium. Concentrations of iridium, the easiest to quantify, were some ~30 times higher than was found in the strata above and below the boundary (see Fig. 2.2). Also there was a much higher concentration of iron pyrite (iron sulfide) in the boundary material, than appeared in the upper and lower layers. (This sulfur-based rock was to have strong significance as will be discussed further in our story).

The researchers, looking elsewhere, also identified the same 65.5 Million year old K-Pg boundary layer in Denmark (some 1400 km distant), which also had a high iridium spike coinciding with the Gubbio data (Fig. 2.2). Furthermore, the Denmark strata showed both similar abrupt changes in the vertical distribution of forams as well as the complete absence of fossils in the thin clay K-Pg boundary, again characteristics matching those of Gubbio. Searching further afield, they found high iridium in K-Pg layers reported in New Zealand—on the other side of the world—18,300 km away.

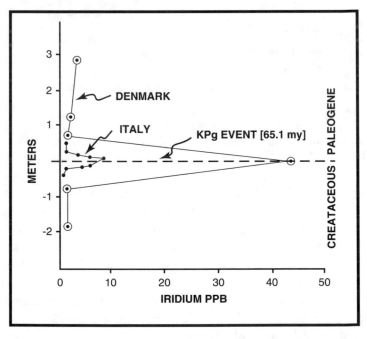

Fig. 2.2 Concentrations of Iridium (parts per billion) as a function of depth (obtained by Alvarez et al.) in K-Pg ancient ocean sediment rocks from Italy and Denmark. The Iridium concentration spikes around the 65.5 million year mark (determined by radioactive dating) and falls to ambient average Earth values below (earlier) and above (later)

What was the significance of the high iridium in the data? A compelling fact was that that iridium (and the other PGE's) was known to occur in much higher concentrations in most stony meteorites than are found in the Earth's crust. This suggested that the high iridium source was of extra-terrestrial origin. Analyses of the isotopic ratios in K-Pg layer samples indicated, however, that this material was still native to our solar system, i.e. unlikely formed from supernova ejecta.

By the process of elimination, similarities of the fossil changes across the globally distributed layers, and the identical age of the boundary samples, lead to the conclusion that the ubiquitous K-Pg layer was formed from the fallout from an impact explosion of a very large carbonaceous stony asteroid—native to our solar system. The extended geographic distribution and the thinness of the K-Pg layer also suggested that the iridium-rich material was spread worldwide over a short time interval, much of it settling out, perhaps within 10 years or so of impact.

On the Size of the Asteroid

An important parameter of an impacting asteroid is its size, which as we have shown in Chap. 1, governs the amount of explosive impact energy and hence is an indicator of its destructiveness to the Earth's living inhabitants. Alvarez et al. obtained a rough estimate of the asteroid's diameter, through four different thought experiments:

1. From several samples of the K-Pg layer an average Ir concentration per unit area of the Earth's surface was estimated. Multiplying this value by the Earth's area yields the amount of Ir delivered by the asteroid that had settled out in the K-Pg layer. Using the known average fraction of Ir in the typical stony-type asteroid, the total mass fraction of fallout forming the K-Pg layer was estimated. However, this left out the Ir within the ejecta that had remained in the atmosphere long after the K-Pg layer was formed. From studies of the Krakatoa eruption, it was estimated that ~20 % of the ejecta's finer particles remained in the atmosphere for a very long period.[6] Adding this amount to the K-Pg layer's volume and using the mean density of 2200 kg/m³ yielded an asteroid diameter of ~6.6 km.

---

[6] Krakatoa was the great Sumatra island volcano eruption in 1883 that caused the largest documented Earth explosion, discussed further in Chap. 3.

2. Estimates of the frequency and the size of Earth orbit-crossing asteroids equal or larger than 10 km in diameter and the craters they produced (see Fig. 1.6), suggests that the mean time between collisions is about 100 million years—consistent with a 65.5 million year-old event. Moreover, the average time between major life extinctions happens to be ~100 million years, further suggesting that a 10 km object would be a reasonable candidate.

3. The K-Pg boundary thicknesses at widely separated sites would yield an approximate volume—if one assumes that the clay material in the K-Pg boundary was the total fallout from the atmosphere with no mixing from the layers above and below. Since asteroid impact models predict that the amount of ejecta thrown into the atmosphere was about 60 times the mass of the asteroid, working backwards, yields the asteroid's diameter as ~7.5 km.

4. The size of the object needed to supply enough fine particulates to be thrown into the atmosphere to shut out sunlight and stop photosynthesis, has been calculated. Referring again to the Krakatoa eruption, the particle concentration required for sunlight attenuation is relative to that of the asteroid impact and suggests its explosion energy was ~1000 times that of Krakatoa. This yields an asteroid-size calculation consistent with those from the other three methods.

The authors concluded, based on these four estimates that the Chicxulub's diameter was ~10 ± 4 km. Using this size, together with a typical meteoric speed of ~25 km/s, it was calculated that the object's impact crater must have been more than 200 km in diameter. As we shall later see, this estimate of the crater size proved amazingly accurate.

Conclusions of the Alvarez Study

The Alvarez paper concluded that it was the asteroid's impact that had created the wide spread extinction in the biosphere at the K-Pg boundary. The impact explosion had propelled a mix of Gulf water, asteroid debris, sea bottom sediments, limestone and other fine rock fragments, into the atmosphere—totaling some 60 times the object's mass. The initial blast pressures, and heat equivalent to 10^8 Mt (~7 billion Hiroshima bombs), instantly destroyed life for hundreds if not thousands of kilometers out from ground zero. Within hours, the daylight turned into night, bringing on severe global cooling, and worse, shutting down photosynthesis on land and oceans. Aerosols and finer particles lingered in the atmosphere and prolonged the cold and darkness likely for years.

The presence of fossil structure and sediments around the K-Pg boundary, and from several other studies, suggested a destruction of 60–80 % of the total Earth's animal and plant life. This included marine and flying reptiles, dinosaurs, larger mammals, most marine invertebrates, microscopic floating animals, and a variety of trees and plants.

The Alvarez paper generated intense interest (including much controversy) so scientists started to look around the planet for a very big late-Cretaceous asteroid impact site. However, as the Alvarez paper suggested, finding the crater would be problematic since, (1) land impact signatures could have long been buried by sedimentation or worn away by erosion, and (2), for a deep ocean impact (more probable) plate tectonics over the eons could have subducted the crater remnants into the mantle.[7]

Found: A Crater

Ironically, it turned out that an earlier observation had been made in the Gulf of Mexico that would eventually support the Alvarez hypothesis. In 1951, the Mexican oil company—Petroleos Mexicanos (PEMEX)—in its search for oil and gas had drilled exploratory wells into the Campeche Bank, the shelf north of the Yucatan Peninsula (Fig. 2.3). One borehole reaching 1300 m below the sea bottom unexpectedly came upon a layer, not of oil or gas, but of fragmented rocks (breccia) containing unusual fused rocks and shocked quartz minerals. Such materials (discussed further in Chap. 3) are often found at sites, which had experienced intense pressure pulses from traumatic explosive events, such as volcanic or atomic blasts. This site was documented as a volcanic dome—an anomalous feature that did not match with the local geology. Over the next 40 years, these data lay unexamined in company files.

In 1978, a PEMEX geophysicist Glen Penfield, while examining airborne magnetic survey data from the Campeche Bank, observed large semi-circular arc-like feature of magnetic anomalies of "extraordinary symmetry" (Penfield 1978). It appeared as part of a ring ~70 km in diameter in the sub—bottom, its east and west ends opening to the south (see Fig. 2.3—insert). Penfield, searching further, then found a map of Yucatan gravity anomaly data, made over the Bank some 10 years earlier, which yielded an even clearer semi-circular feature.[8] This appeared as a deep formation since it was not obvious in the shallow sea-bottom bathymetry. Searching further, Penfield uncovered a later PEMEX gravity study, this time on the Yucatan peninsula, whose gravity

---

[7] This refers to "sea floor spreading" where volcanic magma at the mid-ocean ridges spread over the sea floor and at the continental edges is forced under the plates back into the mantle (the process known as subduction).

[8] This map suggested evidence of an asteroid impact to R. Baltosser, a contractor, but corporate policy prohibited his publishing at that time.

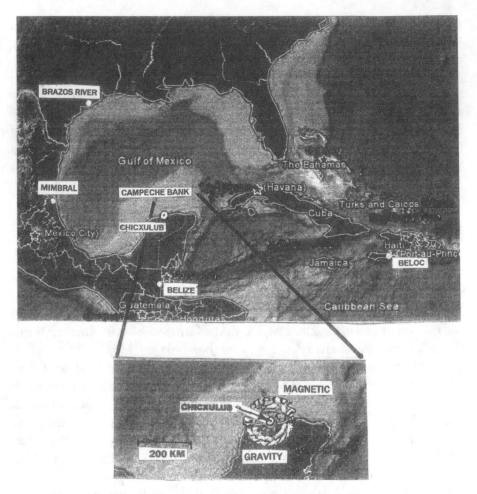

Fig. 2.3 The shallow Campeche Bank, a limestone extension of the Yucatan Peninsula, site of petroleum exploration drillings and the asteroid ground zero (*white circle*) centered at the fishing village of Chicxulub. Insert: The arc formations (upper-magnetic and lower-gravity) anomalies form the ~180 km circular crater

anomalies traced out another semi—circular arc—but whose ends opened to the north. Bingo! The two arcs fit remarkably well together; forming a circular pattern exceeding 180 km diameter and whose center was near the coastal village of Puerto Chicxulub (see Fig. 2.3). The inference here is the magnetic and gravity anomalies suggested that a giant crater had been discovered.[9] It seemed that Alvarez's asteroid (to be) had found a home.[10]

---

[9] Local gravity in the impact site would have been altered, as was the magnetic field. The impact would have compressed or shifted the rock altering local gravity values. Crater patterns such as deep rings would show up as arc-like anomalies in the surface gravity data.

[10] It turned out that this crater, although existing under the sea, was not in a subduction zone such as borders the U.S. west and east coasts, hence it was shielded from sea floor spreading and preserved over the 65.5 million years.

Found: Tektites in Haiti

Fast forward to 1990, when Alan Hildebrand, a graduate student at the University of Arizona, was studying K-Pg strata deposits in an ancient limestone outcrop near the mountain village of Beloc, Haiti ~20 km southwest of Port au Prince, (see Fig. 2.3). This site had been considered to be an ancient volcano by Haitian Professor F. Moras. Hildebrand's analysis of the K-Pg boundary clay, however, revealed randomly distributed fused green glass spherules (Hildebrand 1980). These objects, called tektites (Fig. 2.4) are produced during high-energy events such as violent volcanic eruptions, nuclear blasts, or even from the rare asteroid impacts.

In the heat of the blast, minerals are fused (or even vaporized) as they are thrown out from the impact site. In the atmosphere, they condense and solidify, often forming semi-spherical or teardrop shapes ranging in size from 0.1 to 10 mm that then fell back to Earth. Analysis of the Beloc K-Pg layers also yielded high levels of Iridium. And, interestingly enough, adjacent to K-Pg strata, were large concentrations of ancient coastal marine sediments. Was it possible that these widely spread tektites were formed by ejecta from a massive meteor impact explosion? If so, then the marine sediment materials accompanying the tektites, could have been deposited by turbulent mixing of tsunami waves generated by an asteroid's oceanic impact. Hildebrand, unaware of Penfield's studies, was likely musing, "What happened to the crater"?

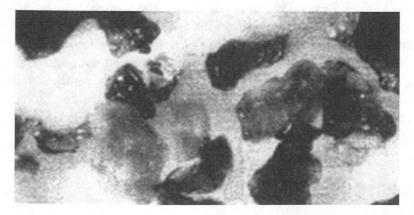

Fig. 2.4 Tektite glass spherules ejected in a molten condition from the Chicxulub impact found in Mimbral, Mexico (Photo by Hildebrand et al. 1991)

A Fortuitous Geology Conference

Later that year, when Hildebrand was presenting his findings at a geology conference in Texas, a Houston Chronicle reporter, Carlos Byars, told him about the 1978 (Penfield et al. 1981) discovery of the Chicxulub crater. Subsequently, the two scientists met and maybe toasted with Margaritas to a "Eureka Moment" as they together assembled pieces of the puzzle. The entire crater with evidence of its global debris field had been found. The Alvarez paper had set the scene and raised the questions. Then, with the Penfield and Hildebrand et al. discoveries, many of the dots had been connected.

Word of the impact crater discovery spread and soon opened the floodgates of further Chicxulub studies. Additional investigations attracted the talents of a diverse group of scientists including: geophysicists, geologists, geochemists, astronomers, biologists, paleontologists—and even cosmologists. Over the next 25 years or so, investigations of Chicxulub, using new scientific and engineering techniques, included the: (1) exploration in the crater using deep seismic refraction profiling studies, (2) borehole drillings from both in the crater area and globally in deep ocean sites. And, (3) further discovery and analyses of worldwide K-Pg boundary outcroppings.

For the remainder of this chapter, we briefly summarize the results of the Chicxulub studies undertaken since the Penfield and Hildebrand interaction in 1990, much of which is based on two comprehensive review papers. The first, published in 2010 by Peter Schulte et al. (with 40 co-authors and 57 references) was entitled "The Chicxulub Asteroid Impact and Mass Extinction at the Cretaceous-Paleogene Boundary." This paper describes the asteroid's impact and its effects on the Earth's environments. It details analyses of deep core and outcrop samples of the K-Pg boundary and examines the effects on mass life extinctions.

The second paper published in 2013, is by S.P.S. Gulick et al. (with five co-authors and 89 references) entitled "Geophysical Characterization of the Chicxulub Impact Crater". This paper concentrates on the geology and has solid review of the geophysics of the Chicxulub crater features at impact and details of the present structure. Also comparison is made of Chicxulub structures with two of the largest Earth impact craters; Sudbury, Canada and Vredefort, South Africa as well as two multi-crater Lunar structures.

Geophysical Gravity and Seismic Data

Following Hildebrand's discovery, further gravity and seismic refraction measurements, both on land and over the sea-bottom, helped to determine shape (morphology) and size of the deep impact formation. An example is a

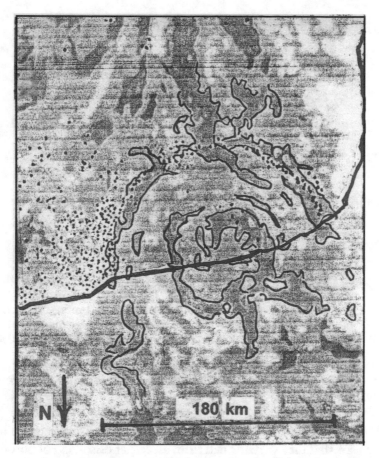

Fig. 2.5 Gradients of gravity anomalies from the deep Chicxulub crater reveal circular patterns of denser rocks created under the great impact heat/pressures centered on the present coastline (*black line*). Black dots on land are water-filled sinkholes (cenotes) probably formed from solution in the porous limestone formed over the crater from millions of years of sedimentation (A. Hildebrand, M Pilkington and M. Connors, 1995)

computer mapping of old gravity data anomalies (Fig. 2.5, shaded areas), was made by the Geological Survey of Canada. The mapping shows at least two dominant concentric craters with diameters of ~80 and ~180 km, centered on the present north coast of the Yucatan Peninsular (black line). Other features suggest a widely distributed debris field.

Along with the gravity patterns, a curious surface phenomenon was found on the landward (south) side of the outer crater. Black dots that were super-imposed on the map posit hundreds of surface limestone sinkholes (cenotes). Their distribution, obtained from both land surveys and by satellite imaging,

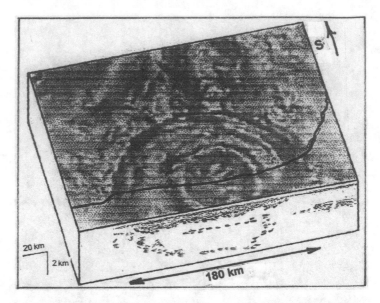

Fig. 2.6   The Chicxulub crater (looking south at about 60°) portrayed by computer analysis of the three dimensional gravity anomalies as underlying multiple crater rings centered on the present coast (*black line*). The cutaway (front face) displays the sub-bottom profile from a marine seismic section ~20 km off shore. Visible is the crater top with down-dropped blocks on its sides and the overlying layers of recent sediments, the same that formed the present Yucatan and Florida Peninsulas (M. Pilkington and R. Hildebrand 2000)

sharply frames the southern rim of the outer crater.[11] Their cause is unclear, but it appears that they were formed from an earlier dissolving of porous limestone accreting along the crater boundary that surrounded the impervious melt material during eons of post impact sedimentation.

Another high-resolution computer contouring analysis of the PEMEX data shows the shaded zones of gravity anomalies (Fig. 2.6). The circular structure of the multiple rings is portrayed magically as if the overlying Gulf water and the upper Yucatan limestone carbonate platform had been swept away.

The front cutaway view in Fig. 2.6 is of a vertical sub-bottom acoustic reflection profile made from a survey ship along an East-West transect.[12] The upper dark area is the ~1–2 km layer of limestone over the crater which formed the Campeche Bank (see Fig. 2.3) which is similar to the Florida and

---

[11] Such formations, found throughout the low carbonate platforms of the Yucatan, Florida, and the Bahamas, are formed when ground water dissolves limestone channels, forming deep wells, caves and even sinkholes.

[12] To do this—as the seismic survey vessel proceeds, acoustical energy is projected downward and reflects off the different bottom layers, arriving back to the ship's hydrophones at different times. This creates a sub-bottom profile map of the various sediment and rock layers along the path (section) of the ship's course.

Bahama Bank platforms. Below, is the basin melt rock layer whose sides, are the wall profiles of tumbled rocks and breccia (fractured minerals) probably from the collapse of the initial transient crater (discussed in Chap. 3).

Most recent surveys confirm the crater as a multi-ring basin, whose outermost or most defining ring exceeds 200 km in diameter with debris fragments out as far as 50 km beyond. This established it as one of the largest impact craters ever-discovered on Earth, Moon, or Mars, Morgan J, et al. (1997); Morgan J and Warner M (1999).

Bore Hole and Crater Stratigraphy

The explorations of the Chicxulub crater and mapping the global distribution of the K-Pg boundary layer greatly benefited from the technologies of the oil industry. Giant floating oilrigs positioned in the Gulf drilled through the ocean floor sediments deep into the bedrock. These bore holes better revealed the 1–2 km of post-impact sedimentary Cenozoic layer of limestone (see Fig. 2.1) composed of fossil foram shells laid down during the ~65.5 million years after impact. Beneath this layer lay the crater basin—a ~50–200 m thick layer of impact melt breccia. These contained shocked minerals including quartz, feldspar, zircon (see Fig. 2.7), which under the Microscope displayed evidence of instant crystalline deformation imposed by the impact pressure waves.

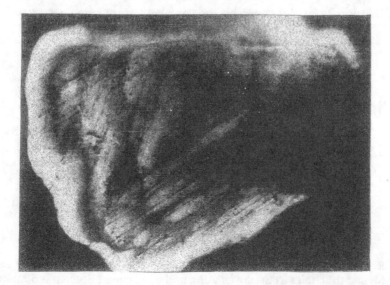

Fig. 2.7 A shocked quartz grain recovered from the breccia mixed melt sheet in a drill hole 50 km from the crater center. The shock impact causes the striations in the crystal planes (photo by Hildebrand et al. 1991)

Beneath the impact breccia is the heart of the crater—the pancake shaped "melt mass" of solidified rock up to 500 m thick and ~80 km in diameter, which, shortly after impact, formed from the molten rocks flooding the crater basin. The potassium-Argon dating, of 65.5 million years of the melt mass samples, was identical to that of the impact glasses found in the worldwide distribution of the K-Pg boundary clays.

Evidence of Ejecta in the K-Pg Boundary

Over the years further PEMEX data were obtained at various distances within and around the crater. Moreover, new core samples were gleaned from some 350 K-Pg boundary global locations—from both land and ocean drill sites (Fig. 2.8).

The K-Pg layer thickness, the character and the size of particle ejecta, and changes in geochemistry, were related to the distance from the crater. Within the crater zone there was more evidence of super-heated mass gravity flows. Within the ~200–500 km of the crater center, breccia layers up to 100 m thick were found mixed with spherules of tektite glass.

Further out, around the ancient Gulf borders, cores yielded layers varying from 1 to 80 m thick and contained a heterogeneous mixture of breccia clastic, fused quartz rock spherules and shocked quartz grains.

At distances of ~1000–5000 km from the crater in the Western Canadian interior and Southeast Caribbean, up to 5 m layers of clastic older rocks along with continued high concentrations of iridium were found. The Marine cores, over 5000 km from the crater boundary layers, were often characterized by

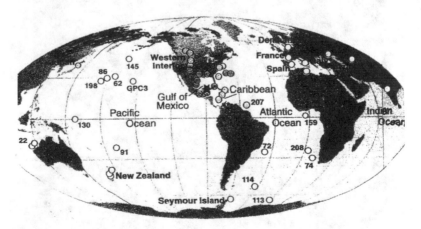

Fig. 2.8 Worldwide locations of K-Pg boundary sites which including outcrop riverbeds and drill holes in the crater zone and deep ocean bottom sediment cores (From Schulte et al. 2010)

a reddish 2–5-mm thick clay that was high in iridium and mixed with finer-grained impact tektites.

Following the original Alvarez K-Pg outcropping sites in Italy, Denmark, and New Zealand (and Hildebrand's findings in Haiti), further outcrops of impact ejecta were uncovered in Belize, North Mexico, Texas (Fig. 2.3), and even in Wyoming. Most of these outcrops were found on river banks where the natural erosion had cut through and exposed layers of geologic history.[13] Examples of the boundary outcroppings are described below.

Arroyal Membral, Northeast Mexico

A dense distribution of tektite glass were found in the K-Pg layers at the Arroyal Membral riverbank in Tamaulipas State, Mexico (Fig. 2.3), some 900 km north-west of the Chicxulub crater. The Iridium rich K-Pg layer, a few cm thick, was underlain by a ~6 m thick layer of sandstone breccia and tektite spherules. Also found in the boundary were tiny diamonds, up to 30 μm in diameter, that may have been produced by the mega impact pressures shocking ambient graphite or possibly from condensation of the vaporized carbon plume.[14]

Brazos River, Texas

Moving north into Falls County, Texas, was found the eroded bank of the Brazos River some 1050 km north-northwest from the crater (Fig. 2.9). The K-Pg boundary displays strata of breccia, in which were found with anomalously high concentrations of iridium, gold, rhodium, arsenic, antimony, and selenium—all typically remnants of the exploding Chicxulub fragments. As was found in Haiti beneath the K-Pg layer, were ancient sea sediments that were possibly deposited by tsunami waves spreading over the shelves and onto the coastal planes in the ancient Gulf.

Dogie Creek, Wyoming

A most unique collection of Chicxulub tektites was uncovered in the K-Pg layer at Dogie Creek, Wyoming (Fig. 2.10) 2800 km north of the crater. A thin (~2 cm) K-Pg layer contained almost perfect ~1 mm spherules. These

---

[13] A most notable example of outcropping is the Grand Canyon, which is but a pair of grossly oversized riverbanks.

[14] We note that tiny diamonds were recovered at Barringer crater (Fig. 1.5). They had probably formed at impact by the mega pressures and heat imposed upon ambient carbonates.

Fig. 2.9 The K-Pg outcropping at Brazos River, Texas. The upper strata contain heated impact breccias (high in iridium) ejecta blasted from the crater, which settled out after impact forming the K-Pg boundary layer. The lower layer includes consolidated sedimentary mudstone (calcareous sandstone) mixed with underlying sediment material all of which was probably transported and deposited inland by giant tsunami waves sloshing around the Gulf (A. Hildebrand 1991)

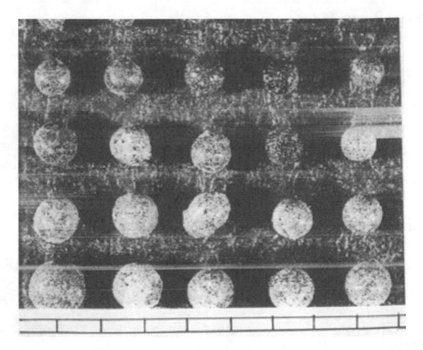

Fig. 2.10 Tiny tektites from the K-Pg boundary at Dogie Creek Wyoming. Apparently condensation of vaporized liquefied mineral glass formed the near perfect millimeter—sized spherules (A. Hildebrand 1991)

tektites, originally composed of fused glass, were pseudo-morphed (replaced by a secondary mineral) roughly akin to the mineral replacement of the wood in petrified forests). The spherical shapes suggest the objects were condensed from superheated vapors that, due to surface tension (like the creation of raindrops), formed spherical droplets and solidified.

A Typical Deep Ocean Core Through the K-Pg Boundary

Deep ocean sediment cores from the Atlantic and Pacific Oceans tell their own story of Chicxulub. The following is a schematic record example of a core from the North Atlantic Ocean (leg 207, some 3000 km SE of the crater) in Fig. 2.11 (Schulte et al. 2010). This core stratigraphy displays a record of events across the boundary located at the 65.5 million year horizon. The 2–3 cm thick K-Pg clay boundary contained tiny spherule-rich tektites 0.1–2 mm in diameter. Immediately below the layer were shocked-created minerals mixed with calcite and quartz grains.

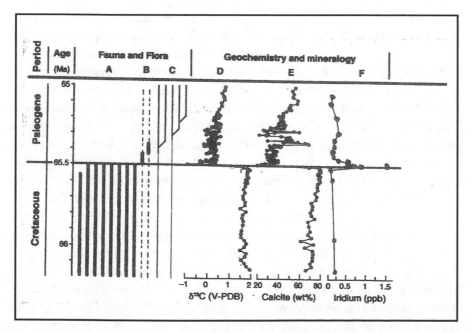

Fig. 2.11 Schematic record of suggested events occurring across the K-Pg boundary correlated from geochemical and mineralogical records from the deep sea North Atlantic Core 207 located in Fig. 2.8 (P. Schulte, et al. 2010)

Graph A suggests the mass extinction of plant species at the Cretaceous boundary while new species (B and C) were formed near the beginning of the following early Paleogene period. Graph D shows the vertical distribution of the quantity "delta" δ^{13} (a reference standard for organic productivity levels) indicating the sharp biotic decrease (i.e., a mass extinction) precisely across the K-Pg boundary.

Profile E likewise shows a sudden drop at the boundary of calcium—based nanoplankton sedimentation (i.e., forams suddenly disappeared). Finally, plot F indicates spiking of iridium associated with the fallout material from the exploding asteroid—clearly replicating the Alvarez data (Fig. 2.2).

Other Attributes of the Crater Target Zone

The Chicxulub multiple ringed craters can be contrasted with that of the relatively small Barringer simple ringed (~1.2 km) bowl—like crater (Fig. 1.5). This effect is also observed on the Moon and Mars where smaller diameter craters display a single rim whereas larger craters tend to have multiple rings. Since Chicxulub's was over 700 million times the energy of the Barringer impact, it likely had more a complicated cratering process. The cause of multiple rings remains unclear. Possibly the initial surface S waves, caused the crust on impact to flex as if liquefied. Then like concentric waves on a pond, they moved out and solidified, forming the rings frozen in time and space.

Another feature of the multi-ringed crater is its asymmetry, suggesting a more oblique angle of attack. Whereas a lower impact angle should produce a more elliptical crater. Thus, energy (and ejecta) should spread out along the semi-major (long) axis. Pope et al. (1996) reports that ejecta accumulations around the outer ring, although quite irregular, suggest that Chicxulub arrived from the southeast at a moderately oblique angle. This would concentrate the impact blast to cause a disproportionate roasting of forests and vegetation toward the lower central United States. Evidence of residual charcoal ash was in fact found in K-Pg layer outcrops northwest of the crater area in Mexico and Texas. However, there was also plenty of carbon available from the impacted deep sediments and limestone.

Moreover, the argument for a lower object impact angle producing an elliptical crater seems in question, since the majority of observed Lunar and Martian multi rimmed craters appear quite circular, even though most of these impactors also probably arrived at random slant angles. Clearly, there is much to learn of the dynamics of cratering from large asteroid impacts.

Two other attributes of the Chicxulub crater were recently established whose evidence could enhance the damaging effects of the impact upon the biosphere.

The Impact Effects of a Deeper Gulf

Closer analyses of crater strata suggested that the ancient Gulf water depths in the target zone were more irregular and extended deeper than had first been assumed (Science Daily 2008). These "deeper depths" were estimated reaching ~1500 m on the north and east sides which increases estimates of the quantity of seawater blown into the atmosphere by ~6.5 times. This amplifies assumptions of the amount of mud and acid-laden fall-out rains returning to the Earth' surface. More importantly, this deeper splash raises the estimates of the size of the initial tsunami wave heights, especially of those formed in the deeper water and headed northeast, toward the Gulf coast and open Atlantic. It is suggested that initial waves could crest at 2–3 thousand meters and advance far inland, drowning coasts of the ancient Gulf and Caribbean. Moreover, it would have greatly increased flooding along the North and South American Continents as well as in Europe and Africa (See more details in Chap. 4).

Effects of High Sulfates in Deep Target Rocks

As previously mentioned, the deep rocks of the impact site contained unusually high concentration of sulfur minerals (e.g., iron sulfide, sulfates, other sulfides, etc.). The impact explosion thus delivered into the atmosphere an estimated 100–500 Gt of superheated sulfur probably as SO_2. Thus, when mixed, with some 35,000 km³ (~35 Gt) of hot Gulf seawater, it created great clouds of aerosols consisting of sulfuric and sulfurous acids. Much of this would condense out and return to Earth as deadly torrential rains, adding greatly to the global acidic-shocking of the biota over the land and in the upper oceans.

Reaffirming the Effects of the "Alvarez Asteroid"

Schulte et al. (2010) and Gulick et al. (2013) re-affirmed the Alvarez claim that the Chicxulub impact was likely the *sole* event—unprecedented in at least 500 million years of Earth history—causing the K-Pg boundary life extinction. It produced a "perfect storm" of multiple (and complex) events that transpired over time intervals spanning from minutes to several 100 years.

The recent studies corroborated the Alvarez findings, providing greater specificity of animal and plant groups that disappeared across the boundary. The list grew to include the non-avian dinosaurs, marine and flying reptiles, ammonites, rudists (spiral shelled mollusks), planktonic forams and calcareous nanofossils. As indicated by Fig. 2.11, new evidence displayed the abrupt alterations of fungi and leaved plants as well as in marine productivity, which likely caused severe changes in the upper food chains. Curiously, the question still remained as to why certain creatures, e.g. crocodiles, snake, and most importantly, many small mammals, survived Chicxulub. More on this is discussed in Chaps. 3 and 5.

Chicxulub's Single Event-Caused Extinction: The Deniers

Some scientists still dispute the Chicxulub event as being the sole, or even the major, cause of the Late Cretaceous extinction. To wit, Schulte et al.'s (2010) Science paper promoting Chicxulub elicited strong negative responses (e.g., Keller et al. 2010). They centered on the Deccan volcanic eruptions which produced massive basaltic lava flows and poisonous gases that percolated through the crust, covering over ~25% of the Indian subcontinent forming the great Deccan plateau.

The Deccan supporters' claim was that the massive eruptions, occurring from ~62 to 68 million years ago, allegedly bracketing the Chicxulub impact, causing severe greenhouse heating along with deadly acidification of land and oceans from toxic atmospheric precipitation. Both of which they claimed strongly, if not principally, contributed to the K-Pg extinctions (Archibald et al. 2010).

There was much confusion over when the Deccan eruptions occurred and how much their effects could contribute to the mass extinction. One question centered on timing—how well the dating of Iridium anomalies, spherule depositions, and tsunami-deposited sediments matched up with the Chicxulub event (Keller et al. 2010) and (Courtillot and Fluteau 2010).

Counter arguments by Schulte et al. (2010) suggested that the Deccan greenhouse gases, only moderately affected climate change, since only ~2 °C warming was estimated at the end of the Cretaceous. Other disagreements concerning the amount of volcanic sulfur injected into the atmosphere that (aside from its greenhouse warming) were needed to produce enough acid rain to cause toxic poisoning of land and the Upper Ocean. Some estimates put the sulfur injection rates at ~0.05–0.5 Gt per year during the ~0.75–1 million-year-long period of extreme Deccan volcanism. But, the Deccan effects would seem no match to those from the Chicxulub impact. It introduced 100–500 Gt of sulfur into the atmosphere within minutes! The question is which is worse for the patient—a long dilute I.V. feeding of poison, or a sudden massive injection?

Over the past several years the "Chicxulubers" appeared to be winning out over the "Deccaners" as to which was the main cause of the K-Pg extinction.

However, new reports from Princeton University, (Stone 2014) suggested that the intense Deccan eruptions started some 250,000 years before Chicxulub and, in fact, lasted ~750,000 years—clearly overlapping the Chicxulub's impact event. A group of paleontologists lead by Gerta Keller will try to determine if the Deccan eruptions were intense enough to have contributed to, or even upstage, Chicxulub's effects on the great extinction.

But wait! To perhaps further complicate this issue—a report in Science by geochemists from the University of California, Berkeley (Renne et al. 2015) in dating of Deccan lavas, found that the eruptions possibly doubled in intensity within ~50,000 years of the asteroid impact. How this happened is still unclear, but it suggests that the massive Chicxulub seismic waves may have modified the Earth crust in the Deccan region allowing increase lava and toxic gas flows to strongly contribute to the mass extinction of the dinosaurs, et al. The controversy continues. Stay tuned.

Based on the data presented in this chapter, we next attempt to construct a scenario describing the sequence of short-term events associated with the Chicxulub's impact.

References

Alvarez L, Alvarez W, Asaro F, Michel H (1980) Extraterrestrial cause for the Cretaceous-Tertiary extinction. Science 208(4448):1095–1108

Archibald J et al (2010) Cretaceous extinctions: multiple causes. Science 328:973

Courtillot V, Fluteau F (2010) Cretaceous extinctions: the volcanic hypotheses. Science 328:973

Gulick S et al (2013) Geophysical characterization of the Chicxulub impact crater. Rev Geophys 51:31–52

Hildebrand AR et al (1995) Size and structure of the Chicxulub crater revealed by horizontal gravity gradients and cenotes. Nature 376(6539):415–417

Hildebrand A, Penfield G et al (1991) Chicxulub crater: a possible Cretaceous/Tertiary boundary impact crater on the Yucatan Peninsula, Mexico. Geology 19(9):867–871. doi:10.1130/0091-7613(1991)019<0867:CCAPCT>2.3.CO;2

Keller G et al (2010) Cretaceous extinctions: evidence overlooked. Science 328:974

Leets L, Judson S (1958) Physical geology. Prentice Hall, Englewood Cliffs

Morgan J et al (1997) Size and morphology of the Chicxulub impact crater. Nature 390(6659):472–476

Morgan J, Warner M (1999) Chicxulub: the third dimension of a multi-ring impact basin. Geology 27(5):407–410

Penfield G et al (1981) Definition of a major igneous zone in the central Yucatan platform with aero-magnetics and gravity. In: Technical program, abstracts and biographies. Society of Exploration Geophysicists 51st annual international meeting, Los Angeles, p 37

Pilkington M, Hildebrand AR (2000) Three-dimensional magnetic imaging of the Chicxulub crater. J Geophys Res 105(B10):23479–23491

Pope K et al (1996) Surface expression of the Chicxulub crater. Geology 24:527–530

Renne P et al (2015) State shift in Deccan volcanism at the Cretaceous-Paleogene boundary, possibly induced by impact. Science 350(6256):76–78

Schulte P et al (2010) The Chicxulub asteroid impact and mass extinction at the Cretaceous-Paleogene boundary. Science 327:1214–1218

Stone R (2014) Back from the dead. Science 346(6215):1281–1283

University of Texas (2008) Seismic images show dinosaur-killing meteor made a bigger splash. Science Daily

3

A Scenario for the Chicxulub Impact and Energies

Abstract After describing the ancient Gulf of Mexico impact site and using the evidence uncovered by the explorations discussed in Chap. 2, we construct a scenario of Chicxulub's fall to Earth and its aftermath—describing the blast that traumatically disrupted the atmosphere, Gulf water, the sea bottom and the crustal rocks. The ejecta that soon returned to Earth formed a scorching heat wave that spread over vast land areas. The totality of the destruction from the impact and its aftermath annihilated 60–80 % of the Earth's animal and plant life, leaving a permanent footprint in its crust.

We consider the physical and geological attributes of Chicxulub's impact site, since they largely determined how the explosion altered global environments. And, finally, we compare several asteroid craters with those from volcanoes comparing their impact (or explosive energies) relative to those of other geophysical and astronomical events.

The Ancient Gulf of Mexico: A Little History

At the time of the asteroid's collision with Earth, in the late Cretaceous period, 65.5 million years ago, the global distribution of the continents were shifted relative to their present positions (see Fig. 3.1). These displacements were, and still are, the result of oceanic and continental tectonic plates—constantly pushed or pulled about by convective forces beneath them in the Earth's upper mantle.

The original version of this chapter was revised. The erratum to this chapter is available at: DOI 10.1007/978-3-319-39487-9_6

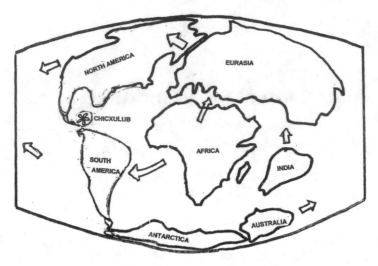

Fig. 3.1 The approximate distribution of the continents in the late Cretaceous ~65 my ago. South America had not joined North America and both were moving westward away from Europe and Africa at a few cm per year. Africa and India was on a collision course with southern Asia while Australia was heading for the South Pacific

South America, having broken away from Africa 60 million years before, was moving west at a few centimeters per year and had not yet connected to North America. Also, the African continent was moving northward toward Europe and India was on its way northeast to a collision with Asia.

Like the global distribution of the continents over the eons, the individual continents' shapes likewise underwent geographic changes (e.g., North America along with the Gulf of Mexico in the deep past had much different structures than exists today). As an example, let us turn the clock back to the mid-Cretaceous period, some 100 million years ago. Studies of ancient sedimentary rocks (e.g., Archibald 1993) revealed a North American continent split into three parts by a huge Y shaped seaway (Fig. 3.2). The Western interior branch extended from the present Central America northward through what are now the central plains and Canada, clear to the Arctic. The western land segment (Laramidia) extended through Alaska, bordered the Pacific Ocean and rested on the Pacific tectonic plate.[1] The eastern branch, the Hudson Seaway, lead up to what is now the Hudson Bay in the northeast, separating most of the Canadian Shield from the landmass of Appalachia.

---

[1] Laramidia subcontinent (coined by Archibald (1993) for Laramie Wyoming) was isolated from Appalachia for a period of 35 million years, during which thriving dinosaur populations rendered this area one of the Earth's richest fossil zones. In contrast, much of the Appalachia fossil beds were destroyed by the more recent Pleistocene glaciation and erosion.

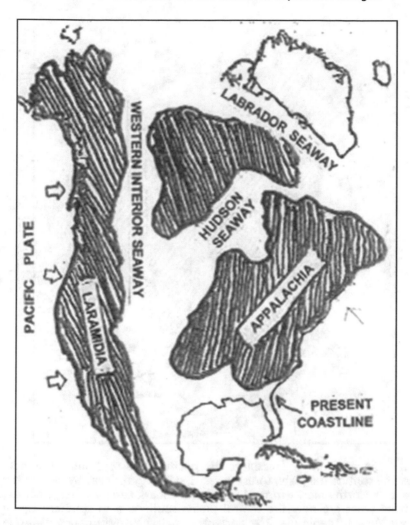

Fig. 3.2 The Mid Cretaceous period ~100 my ago suggests a fragmented North American continent. The Pacific Plate was pushing under the subcontinent of Laramidia forcing it upward, building upon the young Rocky Mountain chain. To the east, two subcontinents were separated by shallow seaways, which joined and extended southward and opening to the western North Atlantic, while the much older Appalachian Mountains were weathering away

This subcontinent was created from the subduction of the Atlantic Plate (which uplifted the Appalachian Mountains 200 million years earlier) and was in the process of eroding away. Meanwhile, the Pacific plate was pushing eastward (as it is today) subducting under Laramidia (see the arrows in Fig. 3.2) causing it to uplift, forming the newer Rocky Mountains.

By the time of Chicxulub, upward displacement and eastward tilting of Laramidia had caused the northern seaway waters to disappear (Fig. 3.3). Left

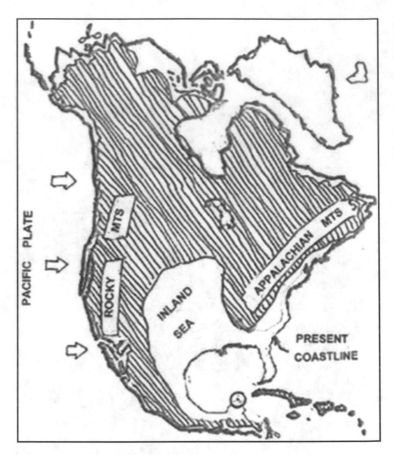

Fig. 3.3 At the time of Chicxulub, the pacific plate continued its subduction under the continental plate, thrust upward, draining the seaways where the late Cretaceous North America formed a single land mass. *Left* was a large bay extending from what are now the central plains southward to southern tip of the ancient Yucatan peninsula. The ancient Appalachian Mountains continued to erode, advancing the eastern shoreline. The Chicxulub impact point (*circle*) suggests the extent of the flooding by its giant tsunami waves over the ancient coasts compared to present coastlines

was a vast shallow "inland sea" (which was actually a bay) of about a quarter of the area of the present United States, its boundary extended some 1200 km northward from today's coast, reaching possibly Iowa and Illinois. Along the ancient Gulf's southwestern side lay a still existing southeast extension of the Rocky Mountains, which turned toward the southeast creating a narrow peninsula that bordered the ancient Gulf waters where the Yucatan Peninsula had yet to be formed.

Chicxulub's impact site (see the circle in Fig. 3.3), in the absence of the Yucatan Peninsular, was over open water with estimated depth varying from

~200 to ~1500 m. East of the impact site, a mountainous island chain had stretched out, forming the Island Arc that bordered the Caribbean Sea. This chain including ancient Cuba, Haiti and the rest of the Antilles islands that curved southward.

Most significant is that the late Cretaceous Gulf of Mexico (i.e., absent today's peninsulas and shelves) encompassed an area over three times that of today's Gulf. It formed a major embayment of the Western Atlantic Ocean—roughly 2500 km across. As we will note in Chap. 4, this ancient Gulf, wide open to the east, greatly enhanced the ability of the giant Chicxulub tsunami to easily radiate its energy into the Atlantic and thence into the world oceans. With today's geography, much of the energy would have been prevented from spreading into the Atlantic Ocean and elsewhere.

The dramatic transformation of the late Cretaceous Gulf into today's configuration underscores the remarkable effects that sea biota can have on the Earth's topography. The warm Cretaceous Gulf was full of sea life, including calcium-based shellfish and micro faunal foraminifera. The remnants of these tiny shelled creatures would continue to filter down onto the sea bottom for the next 65.5 million years, slowly adding to the massive limestone platforms, which today are ~1000–3000 m thick—radically altering the entire Gulf region. This magical transformation, illustrated by Fig. 3.4 displays the coastline extensions that form the present-day Yucatan and Florida peninsulas, the Bahama Islands and adjacent shelves and banks. Much of the land area, that is basically extensions of the shelves, have emerged in part because of the lowering of sea level as polar glaciation increased, (e.g. during the recent Pleistocene ice age ~200,000 years ago).[2] The present adjacent banks (light areas in the figure) make up some of the world's widest coastal shelves.

Having set the target scene, we now imagine the fate of the Chicxulub asteroid as it plunged toward the placid waters of the ancient Gulf of Mexico.

Chicxulub's Atmospheric Entry: The Sonic Boom

Chicxulub was massive. As suggested in Fig. 1.10, its size dwarfed that of the two threatening NEO's: 1950DA and Apophis. With its ~10 km diameter, it had a volume of ~2600 km³—roughly equivalent to some 30,000 rocks of Gibraltar—or perhaps half a Mount Everest. Chicxulub's mass (with an

---

[2] We note that the land areas, i.e. Yucatan, Florida, and the Bahamas with mean altitudes of ~1–5 m will be the first major areas of North America to be submerged as the sea level rises in response to the global warming and subsequent melting of the World's ice caps.

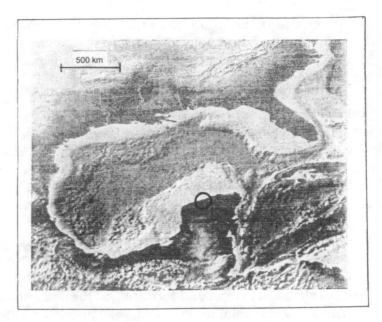

Fig. 3.4 Today's Gulf of Mexico bottom topography showing its extensive coastal shelves (*white*), among the world's widest. The asteroid impact point (*circle*) shows no surface evidence of the 200 km diameter crater buried kilometers below

assumed density of ~3000 kg/m^3), was ~7.8×10^{12} tonnes—equivalent to some 30 million supertankers.[3]

The tumbling asteroid, arriving at ~50 km above the ancient Gulf, entered the thin wispy upper atmosphere at a speed of ~20 km/s. Within milliseconds, the air in front of Chicxulub underwent extreme compression, heating to a bright incandescence. And, as depicted in the Prologue (Fig. P.2), the entire object turned into a glowing white ember as it plunged toward the sea.

Since Chicxulub was travelling many times the speed of sound, the air immediately ahead of it generated a supersonic pressure (P) or shock wave similar to that formed by a speeding bullet or the re-entry of a space capsule (see Fig. 3.5a, b).[4] This supersonic shock wave formed from the air being pushed aside, similar to snow in front of a plow. As demonstrated by the bullet, it falls aside, forming a "bow wave" like the wake of a speedboat. Its direction subtends an angle A_P from the projectile's direction, which adjusts automatically so that the (P) wave moves obliquely to the bullet's path at the

---

[3] Tonne (French) = Metric tonnes (1000 kg) or ~1.1 English tons.

[4] These are longitudinal waves where the particles of the medium (in this case air molecules) vibrate along the wave direction—transmitting energy in pulses. Such P waves can travel through air, water, and solid rock.

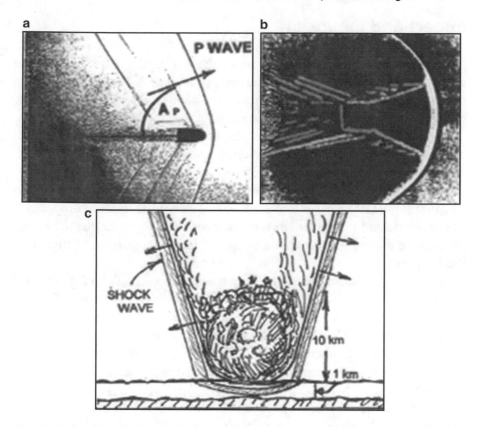

Fig. 3.5 (a) High-speed photo of a bullet creating the supersonic shock wave The P wave forms an angle A_P with projectile. (b) The supersonic shock wave from the Apollo spacecraft re-entry. (c) The Chicxulub's shock wave about to impact the sea surface at ~20 km/s. At this speed of the Angle A_P approaches 90° from the objects path

speed of sound. The faster the projectile, the smaller the Angle A_P. More about P waves is noted in the Seismic Wave section.

Chicxulub (Fig. 3.5c) moved 60–80 times the speed of sound and radiated through the atmosphere almost at 90° from the asteroid's trajectory (see arrows). These P waves produced a mighty sonic boom reverberating across the continents, greatly dwarfing that generated by an SST or Space Shuttle.[5,6] This shockwave was only a hint of things to come, and soon.

---

[5] Folks living in southwest Florida, for years had heard the pronounced sonic boom from a space shuttle gliding overhead for a landing at Cape Canaveral. Try to imagine a sound pulse from the Chicxulub that acted as a giant piston 10 km in diameter and moving at 3 times the Shuttle's speed with 2000 times its cross-sectional area. Of course, with Chicxulub as with the tree falling in the forest, there were no humans to (fortunately) experience it.

[6] While I was on a transatlantic sail to Ireland in the 80s for several days we experienced a daily ear piercing ka-boom. Those of the P wave radiating from the supersonic Concorde SST on its New York run.

Impact and Explosion

After a ~2–3 s passage through the atmosphere, a flaming Chicxulub hit the sea surface like a giant sledgehammer on a steel plate. Note that fast moving objects hit a liquid water surface as though it were a solid; as suggested by the experience of falling while waterskiing at high speed.

The instant compression between the asteroid and dense seawater molecules produced an unimaginable peak pulse pressure, of the order of 10 GPa, equivalent to ~100,000 atm (or ~735 tons per square inch). By comparison, the (ocean) hydrostatic pressure at the deepest point in the Marianas Trench ~11,000 m is ~1100 atm (or 8 tons per square inch). The sudden compression of the atomic and molecular lattices of the rock and water caused their temperatures to skyrocket to tens of thousands of degrees, transforming them into fluid plasma.

A fraction of a second later, the fiery mass, with still much of its forward momentum, plowed into the sea bottom. This instant is suggested in Fig. 3.6. Its kinetic energy was released in a massive explosion in which the water entrained beneath, as well as bottom sediments and bedrock, were consumed in a huge expanding spherical fireball, part of which penetrated some 30 km into the Earth's crust, reaching and distorting the Mohorovicic (MOHO) discontinuity.[7]

At this point in our story, Chicxulub's explosion becomes the Godfather of the most Earthshattering of sequential phenomena of unthinkable intensities. Figure 3.7 is an attempt to summarize the chaotic train of events and their durations. Three principle phenomena were triggered: (1) the generation of powerful **seismic** waves into and along the surface of the Earth's crust, (2) the projection of the megatons of **ejecta** materials into the atmosphere and (3) the production of huge globally spreading **tsunami** waves. In the following we summarize events associated with the seismic waves and ejecta, each of which produced subsets of phenomena, ranging in duration from seconds to thousands of years. The tsunami and its effects are discussed in Chap. 4.

---

[7] A boundary, separating the Earth's continental and oceanic crustal rocks from the hot upper mantle, was discovered by A. Mohorovicic while studying the behavior of earthquake waves. The continental block's thickness is ~30 km, whereas the crust beneath ocean it is ~5 km. It appears that both the (less dense) continental and (more dense) oceanic blocks float upon the semi liquid upper mantle, hence buoyant forces cause the continents to rise higher than the oceanic crust (the sea bottom).

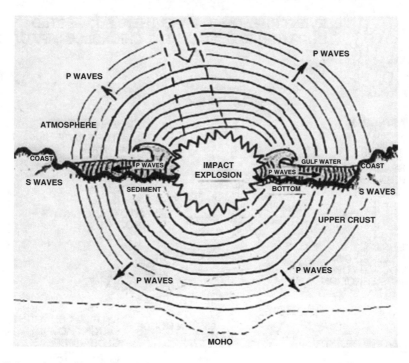

Fig. 3.6 Schematic of the Seismic waves produced at impact. Pressure P waves were generated into the atmosphere, into the shallow Gulf waters, and into the Earth's crust. Shake (interface) S waves were introduced, radiating along the sea bottom, then onto the coasts and inland to the mountains

The Seismic Waves

The impact explosion sent off pulses of seismic waves far exceeding the power of the largest Earthquakes, volcanic eruptions, or nuclear detonations. The first waves were P waves (or pressure waves), similar but much stronger than the asteroid's entry shock waves discussed above.

Pressure wave transmission can be conceptualized as suggested in Fig. 3.8a, where repeated hammer blows (the explosion) on a steel rod (the medium) causes the metallic atoms to collide in sequence, producing longitudinal oscillations which move down the rod creating the rarefactions and condensations of pressure. Oscillations ranging from 20 to 20,000 vibrations per second, the audible range of humans are, for convenience, called sound waves. Energy is transmitted outward, as particles vibrate (oscillate) small fractions of a millimeter. In a single explosion, high-energy P wave oscillations are lumped together forming wave packets. These are suggested as the P waves travel down the rod producing the signals as depicted in Fig. 3.8a.

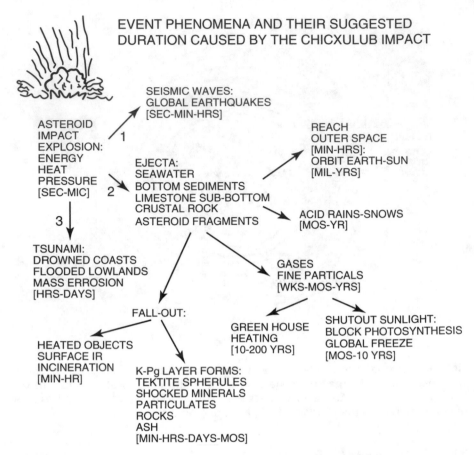

EVENT PHENOMENA AND THEIR SUGGESTED
DURATION CAUSED BY THE CHICXULUB IMPACT

SEISMIC WAVES:
GLOBAL EARTHQUAKES
[SEC-MIN-HRS]

ASTEROID
IMPACT
EXPLOSION:
ENERGY
HEAT
PRESSURE
[SEC-MIC]

REACH
OUTER SPACE
[MIN-HRS]:
ORBIT EARTH-SUN
[MIL-YRS]

EJECTA:
SEAWATER
BOTTOM SEDIMENTS
LIMESTONE SUB-BOTTOM
CRUSTAL ROCK
ASTEROID FRAGMENTS

ACID RAINS-SNOWS
[MOS-YR]

TSUNAMI:
DROWNED COASTS
FLOODED LOWLANDS
MASS ERROSION
[HRS-DAYS]

GASES
FINE PARTICALS
[WKS-MOS-YRS]

FALL-OUT:

GREEN HOUSE
HEATING
[10-200 YRS]

SHUTOUT SUNLIGHT:
BLOCK PHOTOSYNTHESIS
GLOBAL FREEZE
[MOS-10 YRS]

HEATED OBJECTS
SURFACE IR
INCINERATION
[MIN-HR]

K-Pg LAYER FORMS:
TEKTITE SPHERULES
SHOCKED MINERALS
PARTICULATES
ROCKS
ASH
[MIN-HRS-DAYS-MOS]

Fig. 3.7 Event phenomena created by the Chicxulub' impact which produced: (*1*) Seismic Waves, (*2*) Massive Atmosphere Ejecta, and (*3*) Globally radiated tsunami waves. These phenomena in turn produced effects disastrous to the Earth's animal and plant life

The Chicxulub explosion simulated a gigantic hammer blow, sending three-dimensional P waves travelling into the atmosphere, the Gulf seawater, and the Earth's crust. This first group of P waves (like the entry pressure waves but much stronger) radiated into the atmosphere as a sonic boom reverberating around the Earth several times (at 0.6 km/s) at such an intensity to decimate eardrums of land creatures over much of the planet.

The second group of P waves, radiated through the Gulf waters at ~1.5 km/s produced pressures equivalent to billions of anti-submarine depth charges. This caused an instant destruction of aquatic life within hundreds of kilometers from the blast center. Finally, the greatest portion of the P wave energy

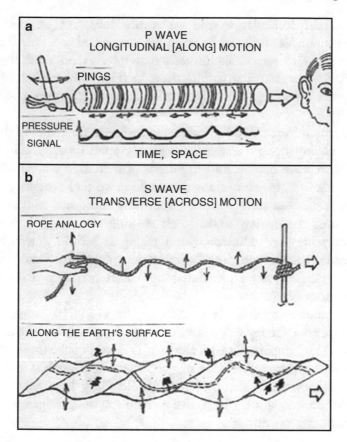

Fig. 3.8 Examples of the generation of P and S waves. (**a**) Pings of the hammer produce longitudinal P waves in a solid metal rod, moving as a train of pressure pulses down the rod. The rod vibrations impact air molecules, which then impact the eardrum producing sound. (**b**) Generation of Shake S waves produce by shaking a rope in a transverse (*cross*) direction. In an Earthquake a sudden thrust causes its surface to move up and down as waves travel outward

blasted into the Earth's crust. Such waves, normally associated with Earthquakes, traveled at ~1.2 km/s as vibrations of rock material.[8]

Following the impact, P waves were the slower, but far more damaging, S (shake) waves (alluded to in the Prologue). S waves moving at ~0.5 km/s (see Fig. 3.6) travel along solid or liquid interfaces, similar to ocean waves (discussed in Chap. 4) where the much of the wave energy is expended trans-

---

[8] It is the study of P waves moving and reflecting within the deep layers in the crust and mantle that provide much information of the Earth's deep interior structure.

versely (i.e., perpendicular to the wave direction). An analogy of such oscillatory motions is produced by snapping a rope tied to a pole (Fig. 3.8b). S waves have much lower oscillation rates than P waves, often less than a cycle per second. S waves from a strong Earthquake (Fig. 3.8b) can exhibit up and down motions exceeding 1 or 2 m causing extreme distortions of the land surface.[9]

The Chicxulub S waves would have emanated from the center of the impact explosion, moving along the seafloor as giant ripples (black line in Fig. 3.6) with 10–20 m wave heights and hundreds of meters in wavelength. These monster oscillations, speeding along the sea floor, would have added energy to the already chaotic sea surface above—spawning additional tsunamis. As the S waves neared the shallow coasts, their shaking would set off underwater turbidity currents (i.e. sediment slides) racing as a fluid down the gentle slopes, scouring the sea bottom and altering the topography. Finally, the waves emerged from the sea, raced inland shaking coastal mountain chains and triggering enormous avalanches.

Asteroid impact models, (e.g., Schulte et al. 2010) suggested that Chicxulub's seismic energy was equivalent to a virtual earthquake of unheard magnitude >11. Compare this with the world's largest quake (in Chile) occurring in 1960, registering 9.6. This suggests that Chicxulub's P and S waves, spreading both over and into the crust, would likely have released stresses along major tectonic faults, which, like a chain reaction, triggered additional major earthquakes, worldwide.

Explosion, Ejecta and Fall-Out

Within seconds after the impact explosion, much of the superheated plasma mixture was blown upward, as if shot from a cosmic cannon. Some 90,000 Gigatons (Gt) of ejecta mixture blasted tens of kilometers into the atmosphere. Larger solid objects which had attained an escape velocity of ~11 km/s, were projected through the outer atmosphere into deep space, having been committed to spend, at least, a few million years orbiting the Sun.

At the same time, slower moving solid ejecta, not having escaped the Earth's gravity, began their downward plunge back to Earth (see Fig. 3.9). The land and oceans within hundreds of kilometers from ground zero were bombarded with debris with masses ranging from kilograms to megatons. As the millions

---

[9] Seismologists, measuring the arrival times of P and S waves on a seismograph and knowing the difference in the speeds of P and S waves, can calculate the distance from the earthquake. With two or more seismographs, location of the quake can be pinpointed.

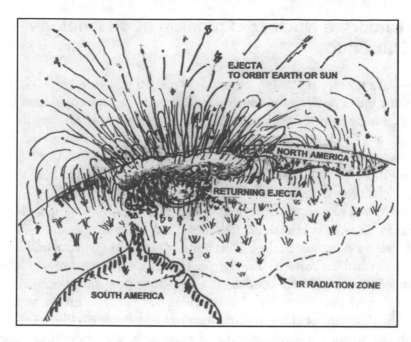

Fig. 3.9 The suggested ejecta field created within minutes after Chicxulub's impact. View from over South America shows the spreading mushroom cloud along with the dense ejecta objects thrown into out space at speeds of several km/s. Solid debris blown into the upper atmosphere reaching the escape velocity of ~11 km/s break away from gravity and proceeded to orbit the sun. Ejecta falling back to Earth, re-heated by atmospheric friction formed a massive infrared wave producing oven-like temperatures over the western hemisphere (*broken line*)

of objects fell back through the denser lower atmosphere, their collective frictional reheating created an infrared pulse that temporarily baked vast areas of the Western Hemisphere raising temperatures to thousands of degrees.[10] Over the following days and weeks, the continuous spreading and settling of the finer particles within the plume initiated a worldwide fallout giving birth of the globally distributed K-Pg layer.

Meanwhile, on the surface of the Gulf, giant tsunami waves created from the splash (as will be discussed in Chap. 4) were moving out and began to spill over the Gulf coasts, through the Caribbean and eventually spreading over much of the Atlantic and Indian oceans.

---

[10] It is estimated that the heat pulse may have lasted for only a few minutes, but it may have been the single most devastating Chicxulub effect upon the Earth's biota.

An Undersea Nuclear Explosion as an Analogy of Chicxulub

As we have noted, envisioning the scenario created by a speeding 10-km rock smashing into a shallow ocean seriously challenges the limits of human imagination. Perhaps a visual analogy (although in miniature) of an asteroid "bottoming out" and exploding in a shallow sea might be suggested by an underwater nuclear blast. This is starkly illustrated by the photo (see Fig. 3.10) taken ~10 s after the "Baker" detonation at Bikini Atoll in the South Pacific on 25/July/1946.[11]

The explosion's energy was estimated at 21 Kilotons (Kt): compared to that of the 14 Kt blast at Hiroshima (Rhodes 1986). The bomb set at ~27 m in depth, blew a giant hole in the lagoon floor, while ejecting several million tons of seawater and reef material into the air. The huge white mushroom cloud (as pictured) was formed from condensed water vapor, propelling outward at several km/s from the searing center of the blast. The dark line on the water suggests the signature of the spreading atmospheric P wave. Above the steam cloud, the mix of hot seawater and reef material is still accelerating upward, forming an edifice ~2–3 km in height and ~600 m across. Beneath the vapor cloud, the vertical cylinder (stem) of heaviest ejecta can be seen starting its downward plunge back into the sea. At the base of the blast, a ~30 m high crest of a mini-tsunami formed among the moored test ships (the largest of which were up to 150 m in length). The shallow water of the lagoon absorbed much of the waves energies, although 2–3 m heights were observed ~7 km seaward from the blast site.

As dramatic as it appears, the energy of the Bikini explosion of ~800,000 (8×10^5) J would have been only comparable to that produced by an impact of a ~20 m diameter asteroid. We leave it to the reader to visualize an ocean splash produced by the 1.1 km (1100 m) 1950DA asteroid, with a mass ~200,000 times greater. Or, the really unthinkable event produced by a 10 km Chicxulub—with mass 100 million times greater.

---

[11] The Baker explosion was part of a nuclear testing program undertaken by the U.S. after WWII.

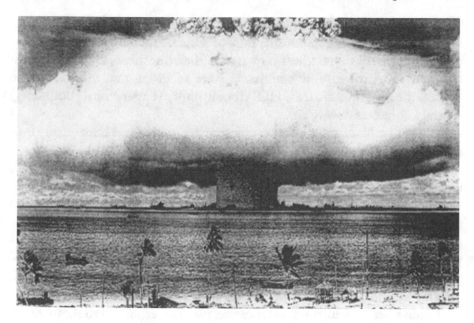

Fig. 3.10 The Bikini underwater nuclear test explosion 10 s after ignition. The white mushroom cloud is the condensate of the lagoon seawater instantly evaporated around the fireball. Above the cloud is the mass of seawater and bottom reef ejecta material climbing skyward. In the foreground the calm breeze will soon be replaced by tornado force winds, while the moored vessels will be tossed about like bathtub toys. US Navy photo

The Chicxulub Crater Formation: A Model

We have seen that the Barringer asteroid impact produced a simple surface crater (see Fig. 1.5) of much smaller and shallower dimensions than those of the Chicxulub crater. Moreover, Chicxulub's more complex crater structure, as revealed in Chap. 2, suggest it was generated by a series of events associated with the asteroid's far greater impact explosion as it interacted with layers of the Earth's crust. A suggested sequence of events producing the Chicxulub crater formation is sketched below as portrayed by Gulick et al. (2013) and Melosh (1989).

The Great Explosion Cavity

As the fiery hot Chicxulub impacted the sea surface of the ancient Gulf, incredible series of traumatic events were about to be played out. The asteroid instantly transferred most of its kinetic energy into heat and pressure—forming the massive explosion centered at the sea bottom. The following chain of events are depicted in Fig. 3.11 (panels a–d), as suggested by Gulick et al. (2013)

Within the few seconds after impact Gigatons of sea water, solid sea sediments, and underlying limestone, part of which, had been turned into liquefied and gaseous ejecta, was blown into the atmosphere (Fig. 3.11a). At the same time, the explosive pressures were so strong that they started to depress the Mohorovicic (MOHO) discontinuity (upper mantle boundary) several kilometers downward.

The portion of seawater not already vaporized, was blown horizontally outward 20–100 km forming a mountainous pileup (Fig. 3.11b). We suggest that it was this giant water annulus whose crest collapse created the Chicxulub tsunami. This is discussed further in Chap. 4 as the "Edgerton effect".

Also portrayed in the figure are S waves moving along the sea bottom (which were depicted in Fig. 3.6). It is suggested that these waves propagated out from the explosion through the surface of the surrounding sediments—which were temporarily liquefied. This would have amplified the S waves into giant ripple-like oscillations, some of which soon thickened and solidified, forming the multiple concentric rings (as evidenced in the Figs. 2.5 and 2.6).[12]

Within several minutes the giant cavity, now had expanded to some 90 km in diameter and 30 km deep. Now its base, forced by hydrostatic back-pressure from the incredibly compressed crustal rock beneath, rebounded upward and formed an expanding central cone (Fig. 3.11b). At the same time the sides of the rim were blown outward expanded the structure to beyond 130 km diameter. Within a few more minutes the now Matterhorn—sized cone and the crater walls, reached several kilometers above the ground level. The structure, being in a highly liquefied state, became hugely unstable and collapsed some 5–10 km downward and inward (Fig. 3.11c).

The collapse of the transient cone and walls, occurring within a span of 10–15 min of impact, resulted in another monumental explosion, producing further cosmic chaos. The circular impact zone roughly 200 km in diameter was a now seething blast furnace (see Fig. 3.11d), a mix of huge fiery limestone melt blocks and smaller shattered rock breccia around the edges which tumbled downward through the sea of molten rock at several thousand degrees. The melt rock from the upper reaches of the transient crater formed giant slurry avalanches which poured downward filling the voids among the slabs of deeper limestone, then spread out over the crater base like a massive griddle cake. This formed the ~20,000 km³ "melt sheet" (the dark zone in

---

[12] A similar but not identical phenomena termed "soil liquefaction" occurred in the 1989 Loma Prieta—San Francisco Earthquake. Sediments around the Bay, shaken by the quake, temporarily liquefied which must have enhanced S wave buildup and propagation, greatly adding to the wide spread structure collapses. In the case of Chicxulub, the sediments and upper rocks must have instantly fused allowing them to better transmit S waves.

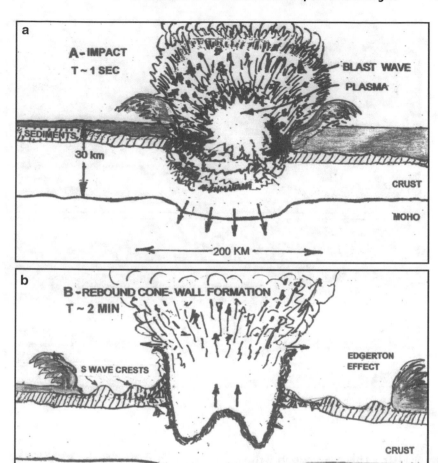

Fig. 3.11 (**a**) The initial explosion forms the hot plasma core while the blast ejecta is blown skyward. The pressure wave blows the Gulf water outward and presses downward on the MOHO boundary 30 km below. (**b**) Within minutes the removal of the Gigatons of ejecta forms the transient crater, while the Gulf water forms a towering outward annulus. It soon collapses, forming the tsunami (the Edgerton Effect). Both the crater base and the depressed Moho start to rebound upward, while S waves radiate over the liquified crust. (**c**) In tens of minutes, the rebound cone of the transient crater has grown to mountainous dimensions, while the MOHO forms a 5 km rebound dimple over its depression. The steep walls of the cone and crator become liquified and collapse over the broken crustal blocks. The S waves, radiating from the crater solidify forming concentric rings. (**d**) Within 24 hours, the crater structure has adjusted: fractured blocks settle and the massive reservoir of hot lava-like fluids form a great melt sheet that spreads out and into the breccia rocks like a "hot fudge Sunday." The boiling seawater is kept at bay for weeks or months by the crater's heat

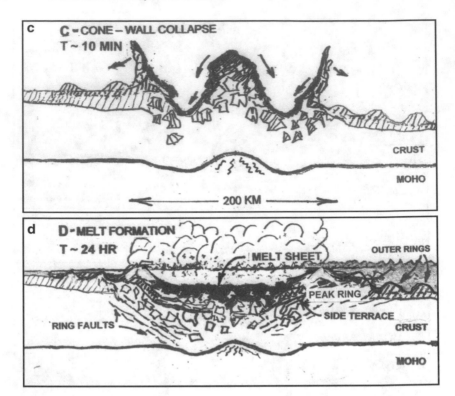

Fig. 3.11 (continued)

Fig. 3.11d). (This was probably the hardened melt rock that the PEMEX1951 borehole penetrated which was erroneously reported as "volcanic-like material).

As the figure depicts, within 24 hr the Gulf water has surrounded the crater covering the outer rings. But over the great melt sheet, which had become a massive heat source, continuously vaporized any seawater spilling into the 10–15 km deep crater. To make things more complicated—if that was possible—the MOHO boundary layer, likely ruptured during the initial impact, may have allowed hot magma to migrate up into the bottom of the transient crater, adding to its heat content. The impact crater edifice now would appear as a monster volcano—crated by an asteroid.

For probably weeks, at the edge of the volcano-like crater, boiling muddy Gulf water heated by the underlying fiery melt sheet, was kept at bay—forming giant billowing vapor clouds. With time, lateral mixing allowed the seawater to gradually invade the crater and start to cool down the simmering caldron. As the seawater poured into the crater's immense volume it produced strong horizontal sieche oscillations.[13] These in turn produced secondary tsunamis which spread over the already drowned coasts of the Gulf and Caribbean.

Another impact phenomenon of interest is the result of an instant and explosive removal of the large volume of seawater and terrigenous material from a single location in a shallow sea. The Alvarez estimate of the asteroid's total ejecta was ~60 times its volume or about 158,000 km³, subsequent impact modeling suggests a more conservative estimate of ~130,000 km³. This is about 5.5% of the volume of the present Gulf, equivalent to nearly 10,000 Lake Superiors or about 0.01% of the world's ocean. This suggests that when the seawater drained into the vast crater it must have caused a temporary drop of sea level perhaps 10–20 m around the Gulf.

The seismic data discussed in Chapter 2, suggested the initial transient crater (see Fig. 3.11b) attained ~30 km in depth. Figure 3.12 compares the size of the Chicxulub initial crater and the partially debris filled crater with the today's deepest ocean depth, the highest mountaintop and the Grand Canyon. It is suggested that the Chicxulub crater even when partially filled with impact blocks and molten magma, and the seawater, was for millennia the most dominant physiographic feature on the planet. Aside from the hydrodynamic disturbances, the creation of this 200 km diameter lens of boiling hot water created major disruptions of oceanographic conditions in the surrounding Gulf. Such are described in Chapter 5.

Today's Crater

The essential features of today's crater are presented in the schematic section Fig. 3.13 [looking west] which depicts the Chicxulub impact's a major disruption of the upper crust. Most obvious is the gross distortion of the deep MOHO boundary which when, slammed with the asteroid's impact P wave was first depressed downward, then rebounded upward formed what looked like the arch of a giant's footprint. Above this is the gross intrusive uplift of crust that filled the great void caused by the explosive removal of Gigatons of ejecta.

At the upper 5-10 km level are strewn the massive debris of rock breccia and huge chunks of Cretaceous blocks, remnants of the collapsed cone and walls from the initial impact crater. Above this zone is the melt rock layer [black] some 100 km in diameter and 3-5 km thick. This is bordered by a peak ring and beyond, 2-3 outer rings. Overlying the breccia and melt rocks

---

[13] Seiches are created in harbors and bays by storms or Earthquakes that cause large sloshing back and forth motions as in a giant bathtub. Such phenomena on a large scale are indistinguishable from tsunamis.

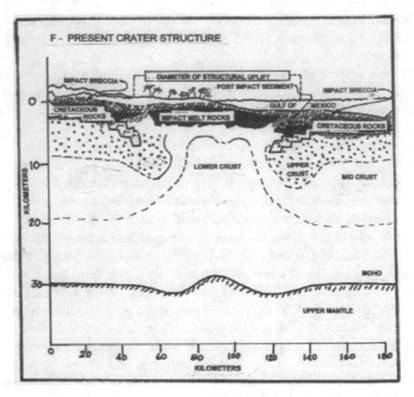

Fig. 3.12 Today, the crater structure and the MOHO distortion have remained as they were after the melt mass cooled and solidified. Over 65.5 million years it has been buried by 2–3 km of calcareous sediments consolidating into limestone rock and topped by soft oolite sediments. The southern half of the 180 km crater lies under the Yucatan peninsula and the rest remains under the Campeche Bank

is the post impact sediment whose deeper layers consolidate into layered limestone. Overlying everything to the north of the epicenter of the crater lays the Campeche Bay bordering the Gulf of Mexico and to the south, the limestone based Yucatan Peninsula.[14]

On Impact and Volcanic Craters

The two most violent explosive phenomena that occur on Earth are (the rare) asteroid impacts and the (much more frequent) volcanic eruptions. Both produce craters whose size and morphology can shed light on their origins. For

---

[14] The Yucatan Peninsula is a flat limestone slab at most 100–200 m above today's sea level. During interglacial periods most of it was under water as was most of Florida. Thus it continued to accrete calcareous sedimentation as did all of the coastal shelves around the Gulf.

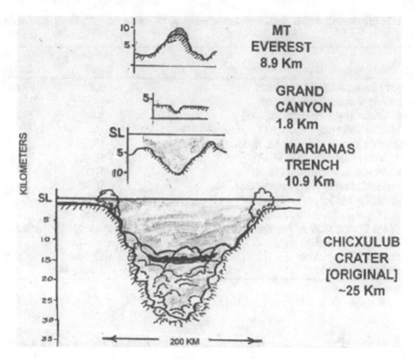

Fig. 3.13 Comparison of the rough dimensions of both the initial and collapsed transient Chicxulub impact crater with those of other present-day Earth geophysical structures

example, Earth craters designated as volcanic are easily identified from their subsurface geology. However, there have been cases of mistaken identity, e.g. the Barringer crater, taken as of volcanic origin since it was difficult to comprehend such a huge impact crater without obvious trace of the projectile. Later, the recovery of small asteroid remnants around the crater proved otherwise. By contrast, the craterless Tunguska was probably comet-like, composed of fragments of brittle stone and ice. Thus, when it entered the Earth's atmosphere, the individual fragments apparently exploded at a height of 20–30 km producing heat and pressure waves that decimated the forests below with only airborne oxide particles and vapors remaining. The recent Russian Chelyabinsk meteorite (see Chap. 2) arrived at a very flat angle, nearly parallel to the Earth's surface, thus dissipating its energy over a wide swath.

It is interesting to compare the attributes of the Earth impact craters with those of volcanic origins. In general, we can assume that, for the more vertical impacting asteroid explosions, the volume of ejecta or of the crater is directly

Table 3.1 Estimated crater volumes associated with volcanic eruptions and asteroid impacts

Event, location, and date	Volume (km³)	Object diameter
Barringer Crater, AZ (~50,000 YBP)	~0.2	50 m
Mt St Helens Volcano, WA (1980)	~1	
Apophis Crater (hypothetical impact)	~3.9	250 m
Mt. Pinatubo Volcano, Luzon, Philippines (1991)	~10	
Krakatoa Volcano, Sunda Straits, Indonesia (1883)	~15	
1950DA Crater (hypothetical impact)	~60	1100 m
Tambora Volcano, Indonesia (1815)	~100	
Yellowstone Caldera, Yellowstone Nat. Pk., CO (~600,000 YBP)	~1000	
Mt Toba Volcano, Sumatra, Indonesia (~74,000 YBP)	~2800	
CHICXULUB Crater, Gulf of Mexico (~65,000,000 YBP)	(~130,000)[a]	10 km
Other natural volumes		
Lake Superior, Great Lakes	~12,240	
Present Gulf of Mexico	~2,400,000	

YBP years before present
[a]Roughly a cube ~50 km (30 mi) on a side

related to the kinetic energy released, i.e. a function of the object's mass and speed—as discussed in Chap. 2. Similar reasoning applies to explosive volcanic eruptions, where estimates of the energy of the eruptions can be approximated by the amount of ejecta thrown out from a caldera.[15]

For perspective, Table 3.1 provides estimates of ejecta volumes of major volcanic eruptions and those of asteroid impact craters. Volume estimates of ancient volcanic and impact craters require rough extrapolations and even estimates of the extent of the millennia of erosion and sedimentation constantly altering the structures. Included are calculated crater volumes of (hypothetical) impacts of the 250 m Apophis and the 1100 m 1950DA objects as well as two notable water volumes given for comparison.

The most well documented volcanic eruption was that of Mount St Helens 1980 dramatically portrayed in I-MAX films. Its explosive energy was estimated at ~5 times that of the 50 m Barringer event. This suggests that Earth impactors such as the much bigger Apophis ("250 m") and the even far more massive 1950DA ("1100 m") asteroids would cause far more catastrophic damage, should they land near population centers.

The direct effects of a volcanic explosion are the production of ejecta, local lava and toxic gas emissions. Longer-term effects are produced by fine particu-

---

[15] For hot volcanic eruptions the heat energy released can be many times that of the kinetic energy of the ejecta, but it's not directly related to the crater formation.

late when projected globally into the atmosphere. This leads to filtering out of sunlight and rapid cooling. Mount St. Helens, Pinatubo, Krakatoa, and Tambora eruptions darkened the skies and led to global cooling (of 2–3 °C that lasted several years. (Today, the most serious effect would be possible damage from volcanic ash being injected into jet engines).

Volcanic craters with estimated volumes greater than 100 km^3 are referred to as "super volcanoes." Notable was the monster Yellowstone explosion occurring some 600,000 years ago. Its deep magma chamber has periodically erupted over several millions of years and is still active—heating the ground water and forming the hot springs and geysers. Needless to say, its seismic activity continues to be monitored.[16]

The largest known volcanic eruption was likely that of the ancient Mount Toba in Sumatra. Its crater forms a beautiful mountain lake 30 km wide and 100 km long. Atmospheric cooling attributed to Toba's giant dust cloud correlates with an abrupt drop of world temperatures of ~10 °C over several years as evidenced by a 74,000 year old core sampling from the Greenland Ice cap.

The Chicxulub approximated crater size clearly dwarfs all others of either asteroid or volcanic origins. This underscores the immensity of disastrous effects Chicxulub must have had upon the entire planet. For perspective, we compare Chicxulub's estimated impact energy release with the energies of other geophysical, astrophysical, or man-made events. Table 3.2 provides an eclectic laundry list of Earthly and extra Earthly energetic events. Listed in Joules, the values suggest order-of-magnitude comparisons. Where appropriate, Richter magnitudes (logarithms, proportional to seismic energy) are provided with some yields also given in Mt. of TNT.

On the lower end of the energy scale lie meteorological events—tornados and hurricanes. The former are associated with intense buildup of kinetic energy of air spinning around an ultra-low-pressure vortex. These "cyclostrophic" winds can exceed 100 m/s (200 knots or 220 mi/hr) and typically these "twisters" occur in horizontal scales of up to ~100–200 hundred meters. By contrast, an Atlantic Ocean hurricane, with winds that may only average 60 m/s (120 knots or 130 mi/hr), but because of their great aerial extent (with typical diameters of ~200 km), far exceeds the kinetic energy of a tornado.[17]

---

[16] Things are changing around Yellowstone Park. The once consistent Old Faithful's eruptions are not as faithful today, started in 1959 to alter their periodicity—likely due to changes in the deep underground heating processes.

[17] Note that generally only 10–15% of a hurricane's energy is kinetic, the rest is heat released by the condensation of its water vapor.

Table 3.2 Energies expended in geophysical, astrophysical and man-made events listed in order of magnitude[b]

Event or phenomena	Magnitude	Yield (Mt)	Energy (J)
Tornado			4.0×10^{10}
Hiroshima Uranium Bomb (1945)		0.015	5.0×10^{13}
Earthquake Haiti (2010)	7.0	0.49	2.0×10^{15}
Hurricane (Atlantic Category 5)			4.0×10^{16}
Barringer Impact (D = 50 m) (50,000 Ya)	7.5	10	1.2×10^{17}
Typhoon (Pacific Category 5)			4.0×10^{17}
Tsar (Russian) Hydrogen Bomb (1955)	8.3	50	4.0×19^{17}
Krakatoa eruption (15 km³ crater) (1883)	8.8	200	8.4×10^{17}
Toba eruption (72,000 BPP)	9.2	800	3.3×10^{18}
Apophis Impact (D = 250 m)[a]			4.5×10^{18}
Earthquake Chile (1960) STRONGEST	9.5	2700	1.1×10^{19}
1950 DA Impact (D = 1.1 km)[a]			2.5×10^{20}
Shoemaker-Levi Jupiter impact (1994)	11.5	10^7	4.0×10^{22}
CHICXULUB (D = 10 km) (65 Mybp)[b]	**12.6**	**10^8**	**1.6×10^{23}**
Typical Collapsed Star Supernova	>35	6×10^{28}	1.0×10^{44}

[a]Hypothetical
[b]For another comparison, the energy of the Chicxulub impact could have supplied the
 US electrical output (based on the year 2000) for some 40 years

Atomic or thermonuclear explosions also fall on the lower end of the energy scale. Their energies are estimated from the mass of fissionable material, blast measurements of pressure, heat, and light records, and even seismic data.[18]

Volcanic eruption energies are variously estimated from crater volume dimensions (see Table 3.1) ejecta mass, and if available, measured seismic and visual data. The Krakatoa eruption, the largest volcanic explosion in recorded history, was much less powerful than the great Toba event 72,000 years ago.

Earthquakes, although not exactly explosions, are included since they represent major energy-releasing phenomena acting over huge areas. Their energies are estimated from multiple seismograph records of P and S wave signals. The strongest recorded earthquake, a 9.5 magnitude that jolted Vivaldia, Chile in 1960 was many magnitudes larger than even the greatest volcanic eruption. It shifted a 1000 km coastal slab, causing a subsidence of several meters and sent a powerful tsunami over the Pacific Ocean. This caused severe damage and loss of life along the US west coast, Hawaii, and the east coast of Japan.

Moving to the asteroid impacts, as we have suggested, the Barringer fall was relatively miniscule compared to the energies that could be released from Apophis or especially 1950 DA, (discussed in Chap. 1). For completeness, we

---

[18]Where the energy = MC², where M is the mass of fissionable material converted to pure energy and C is the speed of light 300 m/μs.

Fig. 3.14 Hubble telescope photo of the surface impacts of Shoemaker-Levi comet fragments upon Jupiter several days after the collision. *Dark spots* of comet debris have widths roughly the Earth's diameter (12,650 km). NASA photo

have included the Jupiter impact of the Shoemaker-Levi comet, one of the rarest of such events observed in our solar system. This was spectacularly recorded by the Hubble telescope (see Fig. 3.14). Several 2–5 km sized comet pieces moving, at 20–30 km/s, ripped into the planet's dense gaseous layers. The impact scars (dark patches) are near Earth-size with an energy release estimated as a Richter magnitude of 11.5, over a thousand times stronger than the great Chilean quake.

The above energy releases fall far short of Chicxulub—clearly the grand-daddy of all known catastrophic Earth events. It produced an equivalent magnitude of a whopping 12.6, exceeding the energy release of Shoemaker-Levi comet collision by almost a factor of ten.

So as to not allow Chicxulub fans to get overly smug about its magnitude, we include the energy estimate of an observed supernova explosion, equivalent to $\sim 10^{21}$ times larger than Chicxulub's. Indeed, that phenomenon produced as much energy in a few minutes as our Sun has produced in its estimated ~ 4.5 billion years of existence. An explosion, indeed!

References

Archibald J (1993) The extinction of the dinosaurs in North America: major extinction of dinosaurs. Geology 24:957–958. doi:10.1130/0091-7613

Christeson G, Gulick S, et al (2006) Moho upwarping beneath the Chicxulub impact crater, GSA, Philadelphia, 22–25 Oct 2006

Gulick S et al (2013) Geophysical characterization of the Chicxulub impact crater. Rev Geophys 51(1):31–52. doi:10.1002/rog.20007

Melosh H (1989) Impact cratering: a geologic process. Oxford University Press, New York

Rhodes R (1986) The making of the atomic bomb. Simon & Schuster, New York

Schulte P et al (2010) The Chicxulub asteroid impact and mass extinction at the cretaceous-paleogene boundary. Science 327:1214–1218

4

The Chicxulub Tsunami

Abstract Having discussed the Chicxulub asteroid's impact and catastrophic explosion that produced Earth-shaking seismic waves and sent Gigatons of ejecta into the atmosphere, leaving a massive crater; we now turn to the third major Chicxulub phenomena—the tsunami. This chapter, after defining three modes of tsunami generation, presents a short "primer" on the nature of tsunami waves, including a description of their most important property, refraction. Examples are presented of historical wave and tsunami events that demonstrate the behavior of refraction and run-up. Two models of tsunami waves are also examined— one created by a volcanic crater collapse in the Canary Islands and the other by a North Atlantic Ocean impact due to an NEO asteroid. Lastly, the Chicxulub impact tsunami formation is portrayed, wherein we track its relentless spreading waves into the ancient Gulf of Mexico thence, into the world's oceans and finally dissipating their energies at the shorelines.

Tsunami Generation: The Three Modes

Most tsunamis result from undersea earthquakes, less common are those caused by volcanic eruptions or Earth slides, and even less frequently, from asteroid ocean impacts. Each of these events that can produce traumatic, even sometimes explosive, vertical displacements of immense volumes of seawater, the collapse of which due to gravity, produces the tsunami.

The original version of this chapter was revised. The erratum to this chapter is available at:
DOI 10.1007/978-3-319-39487-9_6

© Springer International Publishing Switzerland 2017
D. Shonting, C. Ezrailson, *Chicxulub: The Impact and Tsunami*,
Springer Praxis Books, DOI 10.1007/978-3-319-39487-9_4

Earthquake tsunamis are often created from the slip fault release of an oceanic plate thrusting beneath a continental plate (Fig. 4.1a). Sometimes, the subducting plate becomes stuck, resulting in the buildup of great stresses over years, centuries, or even millennia. If the plate suddenly snaps free—much like the sudden release of a jammed cellar door—an earthquake-creating jolt occurs. This causes an instant vertical sea-bottom displacement, typically of ~1–2 m which then produces a corresponding rise or fall of the sea surface. The resulting disturbance, whose area can be thousands of square kilometers,

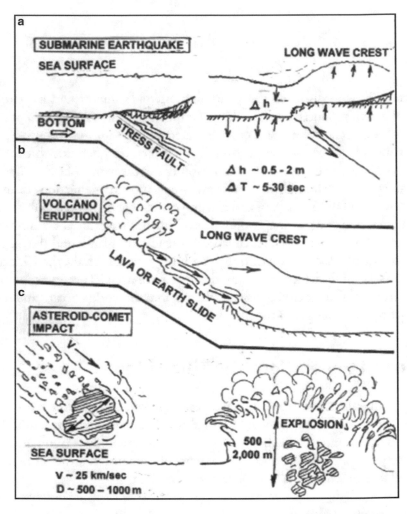

Fig. 4.1 Modes of tsunami generation: **(a)** Release of stress fault causing Earthquake which displaces sea bottom. **(b)** Release of volcanic lava or natural Earth slide into the ocean. **(c)** Asteroid splashing into the sea

rapidly transforms into a series of perfectly formed concentric and ultra-long tsunami waves.

A second mode of tsunami generation is shown (see Fig. 4.1b) as associated with an undersea volcano eruption or an Earth slide. This can send many tens or hundreds of cubic kilometers of rock or lava into the sea, displacing a huge mountain of water that then collapses, forming tsunami waves. A model of a volcanic slide event is described later in this chapter.

Finally, we describe the tsunami created by an asteroid impact into the ocean (Fig. 4.1c). Depending on its size and its kinetic energy, as discussed in Chap. 1, this can produce the most powerful tsunamis ever conceived.

Once generated, tsunamis, governed by simple hydrodynamic laws, radiate away from their source region. They exhibit quite predictable behavior even as they near the coast, where they undergo extreme changes. But, before we examine some examples of historic tsunami events, let us back up and review the physical attributes of these unique waves.

A Primer on Tsunamis: Wave Basics

In the open ocean, tsunami waves occur as a series of very long parallel crests (see Fig. 4.2a) possessing the following attributes:

1. Wave Height (H)—the vertical distance from the trough to crest, see (Fig. 4.2b)
2. Amplitude (A) = (H/2)—the height above or below the mean water level
3. Wavelength (L)—the distance between crests, see (Fig. 4.2b)
4. Period (T)—is also 1/F, the time interval for consecutive crests to pass a fixed point. For a group of waves, T tends to remain constant—irrespective of change in H or L, see (Fig. 4.2c)
5. Frequency (F) is (1/T) or how many cycles occur in unit time.[1]

Ideally, the above parameters apply to smooth sinusoids as are generated in a wave tank or on a laboratory oscilloscope. They are, however, also useful when applied to a field of real ocean waves, whose parameters can be estimated as "eyeball averages" or even better, when they are obtained from statistics of direct measurements.[2]

---

[1] The common example is the frequency of household electrical current, 60 cycles/s or 60 Hertz (Hz).

[2] Ocean waves are detected by several techniques: e.g., spar buoys or subsurface pressure sensors that detect the rise and fall of passing waves. The data analyses require sophisticated statistical techniques.

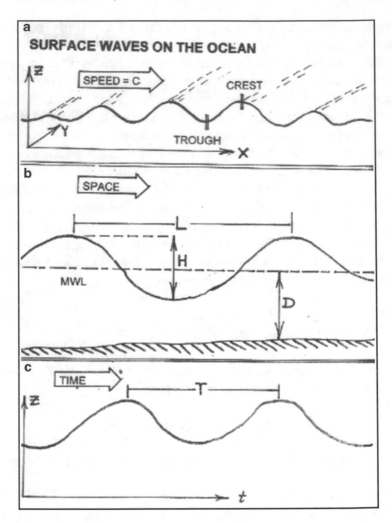

Fig. 4.2 Parameters of surface waves in the ocean (a) a "parade" of surface waves (b) dimensional parameters (c) wave period

Other important parameters include the water depth (D) beneath the waves and the steepness or slope suggested by the ratio H/L. Large storm waves are the steepest with H/L values of 1/20–1/50.[3] By contrast, tsunamis, because of their huge wavelengths and their initially small (~0.5–2 m) wave heights, have vanishingly small slopes e.g., from ~1/10,000 to 1/100,000 and hence, are invisible to the sailor's eye in the open sea. Upon nearing the coast—the L's grow smaller while the H's grow larger—much larger.

---

[3] It is combination of extreme wave steepness and height that produces the unpleasant—and sometimes dangerous pitching and rolling of the ships at sea.

Another important wave attribute is its energy. Recall that in Chap. 1, we defined the kinetic energy (KE) of a rock about to hit the water (Eq. (1.2)). Models suggest that during impact-splashes 10–15 % of the KE is transferred into the surface waves (which have both kinetic and potential energy). The total energy (KE + PE) per unit area of a wave train (Fig. 4.2a) is approximated by:

$$E \sim K \cdot A^2 = 1/4K \cdot H^2 \left(\text{since} A = H/2\right) \tag{4.1}$$

The factor K relates to the seawater density and gravity—both assumed constant. In short, wave energy is proportional the square of its Amplitude (A^2) or height (H^2).[4] Thus, tsunamis coming on shore having similar wavelengths, but heights 2–5 times of wind waves have energies 4–25 times greater. Furthermore, since the total energy per wave is also proportional to wavelength L, this again suggests why tsunami energies are far greater than storm waves since their wavelengths are generally 10–100 times larger.

During wind wave generation, the waves are "forced" by friction or drag on the sea surface. For storm waves to develop, the forcing action (wind stress) must act over many hours or days. This allows hurricanes, and their accompanying waves, to be forecast many hours or days ahead. By contrast, tsunamis are mostly formed within minutes by (generally unannounced) seismic quakes or explosive volcanic events.

When tsunami (also large storm waves) escape their area of generation (usually with maximum heights or energy) they become "free waves." Then, their speed and steering behavior is controlled by the size of their wavelengths L compared to the water depth D.

Free ocean waves occur in two groups. The first are waves whose L values are smaller than the D (i.e., the ratio L/D < 1) and are defined as "short" water waves (Fig. 4.3a). They are also called "deep" water waves because away from the shore most ocean depths are much greater than the surface wavelengths—even of storm waves.[5]

Short waves move at a speed approximated by:

$$C_{SW} = \sqrt{\frac{gL}{2\pi}}. \tag{4.2}$$

---

[4] This is true in general, e.g., for light or sound waves.

[5] Note the largest storm wavelengths are rarely greater than ~300 m (900 ft) whereas the average depth of the world oceans is nearly 4000 m (~2.5 mi).

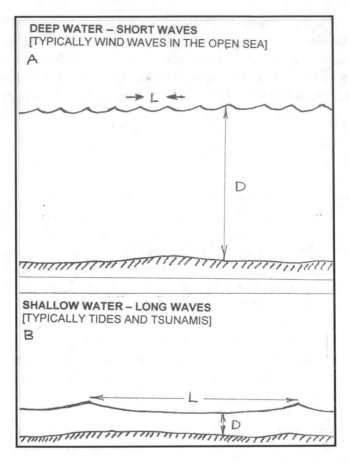

Fig. 4.3 Characteristics of **(a)** deep water (*short*) waves and **(b)** shallow water (*long*) waves

Gravity *g* appears because it exerts a downward pull or "restoring force" on the wave crests, and their fluidity makes them naturally oscillate as the wave profile moves forward.[6] The absence of D in the equation means that short waves move independently of depth, their motions being confined to the surface layer. Thus, the longer their wavelengths—the faster they travel.

Waves in the second group (which includes tsunamis) are "long" or "shallow" waves (Fig. 4.3b). Their L's are greater (usually much greater) than D and their speeds are approximated by:

$$C_{LW} = \sqrt{gD} \qquad (4.3)$$

---

[6] The Greek π appears because model approximates a *sine wave* (from trigonometry fame) that moves in cycles of 2π radians (or 360°): π is about 3.14—the ratio of the circumference of a circle to its diameter.

Table 4.1 Approximate wave speeds for short and long ocean waves

	Wavelength (m)	Depth (m)	Speed (m/s)
Short (wind) waves	20 (60 ft)		5.6 (13 mi/hr)
	100 (300 ft)		12.6 (29 mi/hr)
	200 (600 ft)		20.5 (47 mi/hr)
Long (tsunami) waves		100 (305 ft)	32.0 (74 mi/hr)
		1000 (3280 ft)	100 (232 mi/hr)
		4000 (13,120 ft)	200 (465 mi/hr)
		6000 (19,680 ft)	245 (568 mi/hr)

The g again appears because long waves are also affected by gravity's restoring force. Since C_{LW} contains D, these water motions, unlike those of short waves, extend over the entire water column, i.e., they "feel" or exert a drag on the bottom. The deeper the water, the less the drag; hence, the faster they travel.[7] This bottom effect on long waves gives rise to the important phenomena of refraction, discussed below.

Speeds of short and long waves are given in Table 4.1. The short, e.g. with L ~ 20–30 m and usual heights H ~ 0.5–1.0 m, occur in moderate winds such as afternoon coastal sea breezes. Longer 100–200 m storm-generated waves with H ~ 5–10 m move faster and will easily escape their region of generation.[8]

Clearly, tsunamis with their ultra-long wavelengths are the "greyhounds of the sea"—moving at speeds exceeding those of jet aircraft, far outracing the largest storm waves. This is a deadly attribute since tsunamis can cross entire oceans in a few hours, minimizing the warning time for coastal inhabitants. Even worse, most tsunamis result from slip fault quakes located at the edges of continents, so waves can hit those nearby shores, unannounced, with high energies, often within tens of minutes.

Refraction: Navigator of Tsunamis

The most important property of tsunamis is that of the dramatic change they undergo as they move from deep oceans to shallower waters. This phenomena is called "refraction," defined in the Random House Dictionary as "a change in the direction of a ray of light, sound, heat or the like in

---

[7] Of course there are "intermediate waves" where L/D ~ 1 whose speed is effected by both L and D.

[8] Most storms or hurricanes move (advect) rather slowly, from 5 to 10 m/s (10–20 knots). So 100 m (or larger) wavelength waves easily outrun the storms that generated them and hence, stop growing. Otherwise if the waves traveled slower than the storms generating them they would keep building up and the oceans would be full of monster waves.

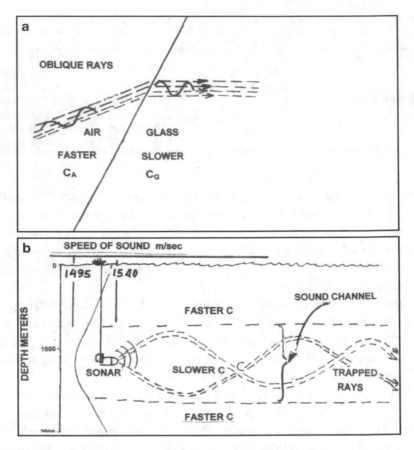

Fig. 4.4 (a) Refraction of light waves as they enter from air to glass. (b) Refraction of sound waves released from Sonar, which constantly bend toward slower sound speed and become trapped in the Ocean Sound Channel

passing obliquely from one medium to another in which its speed is different." Refraction is demonstrated by a simple science experiment consisting of the passing of a beam of light from air into glass at an oblique angle (Fig. 4.4a). The light rays, which move slower in glass than air, immediately change direction or "refract"—bend toward the slower medium.[9] Rays hitting the glass non-obliquely, i.e. perpendicular to its surface, still change speed but not direction and so do not bend.

Not unlike light waves, tsunamis moving as long ocean waves also refract. Consider a train of long waves, moving over three different bottom conditions

---

[9] Sound rays in the ocean behave in the same way. The speed of sound in the deep ocean has a minimum around 1500 m, thus forward rays (like from a submarine) always will bend toward the minimum speed zone, i.e., the center lime of the channel (Fig. 4.4b). Hence they oscillate above and below the depth of the minimum speed, being trapped in the sound channel. This effect can shield a submarine's engine noise from surface ships sonar detection.

illustrated in Fig. 4.5. For waves over a constant bottom depth (Fig. 4.5a) their direction, speed, wavelengths, and heights remain unchanged.[10] In the second case (Fig. 4.5b), waves approach a beach directly, not obliquely, but with an up-sloping bottom (shown by parallel isobaths).[11] Although their direction is fixed, the waves undergo four changes.

1. Their speed slows down because of decreasing depth (obeying Eq. (4.3))
2. The wavelengths L decrease
3. Their wave heights H increase
4. The wave crests steepen and finally break up as turbulent surf

What could explain these changes especially the increase in H? Risking oversimplification of the hydrodynamics, consider the following: waves moving into shallower water slow down, which causes them to cover less distance over the period T which remains ~ constant). The L values decrease; effectively squeezing the wave profile between crests, so as D decreases, the wave motions become more confined. Since the seawater between crests cannot compress nor expand sideways, the crests must grow higher and steeper. Furthermore, this causes the outward pressures on the sides of the crests (due to the growing weight of the water) to soon exceed the surface tension causing the waves to simply collapse or break (like a liquid house of cards) forming turbulent breakers, bubbles, and a little heat.

The third (and most general) case of waves coming onshore occurs when crests intersect the parallel shoaling isobaths at some acute angle A_0 (less than 90° as shown in Fig. 4.5c). The decreasing depth forces the waves nearest shore to slow down sooner than those farther out (in deeper water). This effect, refraction, forces the crests to turn, like the spokes of a Ferris wheel, to steer toward the shallower depths, forcing the wave direction to become more perpendicular to the beach.[12] This reflects the behavior of all tsunamis in the world's oceans: they always meander toward shallower water while growing in height.

Refraction of long ocean swells (like miniature tsunamis) is shown in Fig. 4.6 where wave fronts approach a California coast at an angle A_0 ~ 75° with wavelengths of ~60 m. Nearing shore, the crests turn more parallel to the beach

---

[10] This is an approximation. Real waves slowly lose energy (diminish in wave heights) as they travel, mainly due to horizontal spreading over the ocean and from viscous effects in the water motions, producing a small amount of heat. We neglect these effects.

[11] Isobaths—lines of constant depth displaying the bottom topography, just as "isobars" display lines of constant pressure on a weather map.

[12] Since the ocean depth changes are not abrupt like the air-glass interfaces described above, the wave changes direction more gradually.

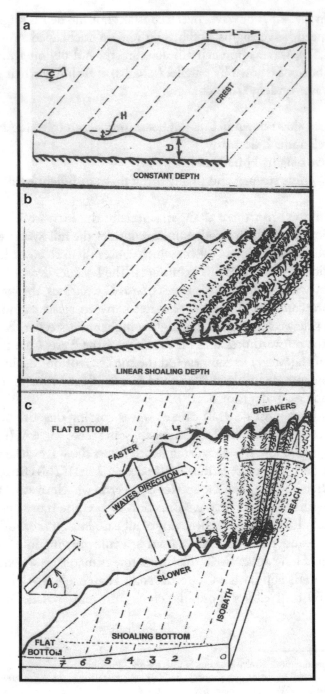

Fig. 4.5 (a) Long Waves traveling in water of constant depth. (b) Long waves moving onto a beach of constant sloping depth and perpendicular to the shore line, and (c) Long waves moving onto a constant sloping beach at an Angle *A* (less than 90°) causing refraction

Fig. 4.6 Refraction of long swells breaking over a beach in California (From Munk and Taylor 1947)

while their wavelengths decrease, forming turbulent breakers. The distortion in the wave fronts (right center) reveals an anomalous local shoaling.[13]

Another important phenomenon exhibited by tsunami waves is that of dispersion. When a tsunami (or even a group of storm waves) is generated, its members are not identical, but each has a slightly different height, wavelength, period or frequency. The group or "spectrum" of these waves move with an average speed determined by the depth (Eq. (4.3)). However, waves with slightly longer wavelengths (and slightly lower frequencies) move a little faster than those with slightly shorter wavelengths (and higher frequencies). In the open ocean, after time, the waves tend to sort themselves out (disperse) with those with slightly longer wavelengths leading the pack.[14] This effect, along with the natural increase in heights in shoaling water, presents

---

[13] In World War ll aerial photos of approaching waves on Pacific island beaches, planned for amphibious landings, were used to indicate dangerous and otherwise invisible shoals to be avoided by landing craft.

[14] While testing instruments aboard a Woods Hole Research Vessel east of Boston, in the afternoon, the seas were calm but we noticed the buildup of long (~100–150 m) swells 2–3 m high from the southeast. These were eerie "forerunners" produced by a hurricane churning toward New England. The effect of dispersion caused the longest of the storm waves to outrace the rest as well as the approaching storm itself. Soon the Captain, heeding the forecasts and the forerunners, headed the ship back to Boston, while fortunately the hurricane passed by offshore.

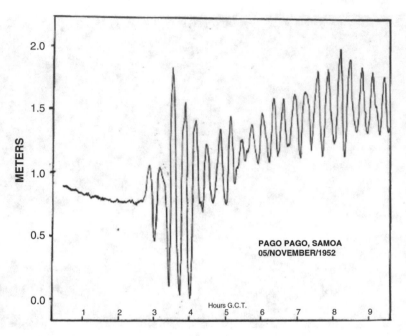

Fig. 4.7 Tide gauge record of a tsunami making landfall in Samoa, south Pacific

all bad news, since it causes the longest and usually the largest (and hence most damaging) tsunami waves to hit the shore first. This effect on a tsunami produced by a quake off the Kamchatka Peninsula is seen in Fig. 4.7, which, having traveled over 5000 km, arrived at a tide gauge at Pago Pago, Samoa, South Pacific. The initial periods of about 20–25 min slowly decrease to 15–20 min. (The smaller waves beginning at 0415 are likely reflection effects around the coast.)

Tsunamis that are born as long waves (with L>D) refract even in the deepest oceans.[15] For example, when waves crossing a 5000 m depth ocean basin move onto a 2000 m mid ocean ridge, it acts as a wave guide, forcing the waves to slow, alter direction and follow the (relatively) shallower sinuous peaks of the ridge. This effect tends to keep tsunamis longer in the open ocean, in effect extending their life span, until, finally meeting their demise upon the shores of the continents.

A striking example of refractive focusing of a giant tsunami as it rampaged over the world ocean was provided by the Indonesian earthquake tsunami of

---

[15] The mean depth of the world ocean is ~4000 m (while the greatest depth in the Marianas Trench is ~10,800 m.) Thus, mostly everywhere in the oceans tsunamis with wavelengths >3000–4000, will refract as long waves.

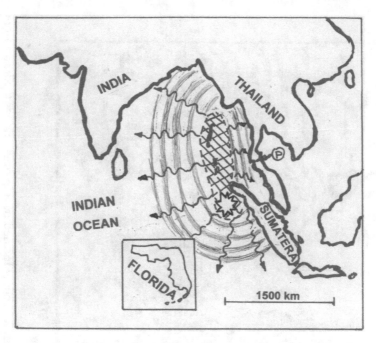

Fig. 4.8 Schematic of the Indonesian Earthquake-tsunami suggesting the tsunami waves radiating from the Sea hatched area Andaman instead of Andean where the *bottom* was severely jolted (*Insert*—The state of Florida, U.S.A. to scale)

26/Dec/2004. This quake, one of the most powerful ever recorded had a magnitude of ~9.1. Its epicenter (the star in Fig. 4.8) was just west of the Sumatra coast. The hatched area was jolted as the Indian Plate thrust under the Burma Plate, suddenly lifting the floor of the Andaman Sea and the Andaman Islands. The area almost twice that of Florida (see Fig. 4.8 insert), experienced vertical displacements of 2–5 m and horizontal shifts exceeding ~20 m.[16] The resulting tsunami radiated huge wave energies westward into the Indian Ocean and eastward directly to the nearby coasts of Thailand and Sumatra.

The tracking of this tsunami over the mid-ocean ridges was dramatically illustrated by a computer model of Dr. Vasily Titov et al. of the NOAA Marine Lab in Seattle, WA (Titov et al. 2005). The model (Fig. 4.9 drawn from Titov et al.'s map) simulated the progress of tsunami wave crests over the globe at hourly spacing. The shaded zones indicate the concentrated energy refracting over the ocean ridges, where the wave heights ranged up to ~1.5 m (most mid ocean tsunami heights average 0.5 m). Heights increased 10–15-fold when

---

[16] Several islands in the Andaman Sea had noticeable increase in waterfront real estate.

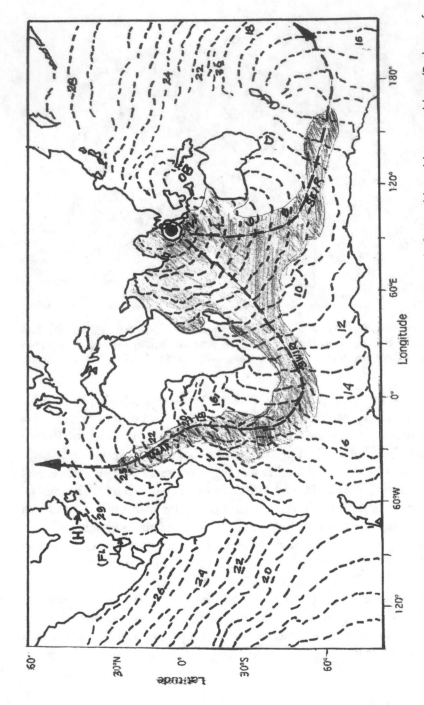

Fig. 4.9 The Indonesian tsunami waves refracting over the globe being strongly focused by mid ocean ridges (Redrawn from Titov et al. 2005)

coming ashore—especially on the Indian and nearby Indonesian-Thailand coasts, causing disastrous flooding.[17]

Waves radiating from the epicenter in the Eastern Indian Ocean (black circle) slavishly follow the axes of the mid ocean ridges (dark broken lines), steering around Africa and then northward, following the Mid Atlantic Ridge. Remarkably, waves registered 20–30 cm on tide gauges halfway around the globe off Florida and Halifax, Nova Scotia. Meanwhile, southeastward moving waves crossed the southern Indian Ocean—between Antarctica and Australia, and then headed up into the Pacific Ocean where they were still detected on tide gauges along the California coast.

Tsunami Waves Coming Ashore: Run-up

When tsunami waves ultimately arrive at coasts, their behavior is affected by how fast (or how steep) the shoaling occurs. When large waves come ashore, water is temporarily carried onto the beach. Oceanographers refer to the term "run-up": an indicator of the waves invasiveness. The amount of flooding is dependent upon the heights, and periods of the oncoming waves. The run-up distance (Fig. 4.11) is the extent the water moves inland horizontally from the mean sea-level boundary. The run-up height is the vertical distance it reaches above mean sea level (MSL). The wave period T is a limiter of the run-up distance since it defines the interval for the wave to advance shoreward and the retreat seaward. Equally important is the coastal topography, affecting the heights of the waves reaching shore.

This effect is demonstrated in Fig. 4.10a when tsunamis moved onto a shallow but broad shelf (say ~200 km wide as occurs off the coast of SW Florida). As their speed decreases and their heights increase, they form large offshore breakers steadily dissipating energy as they cross the shelf. When they finally arrive on shore, much of their energy—and heights have been depleted.[18]

---

[17] Computer predictions are used as alerts and are sent out to coastal sites through a worldwide early warning system. If no dangerous waves appear, the warning is then cancelled. The Pacific has an extensive international tsunami warning system. However, the Indian Ocean was not well prepared and warning to millions of coastal inhabitants was unavailable, resulting in the loss of some 180,000 lives.

[18] The first view of these waves on a Florida Gulf Coast beach was while on my sailboat (Tartan 30) crossing Florida from the Atlantic Coast. The Gulf waves were like those on a lake, with a total absence of large breakers, which are ubiquitous along the Atlantic coast beaches, which often have deeper water directly offshore. The small Gulf waves were the leftovers from the larger wind waves far off shore, having greatly dissipated over the ~200 km shallow shelf, off the coast of Naples, one of the world's largest.

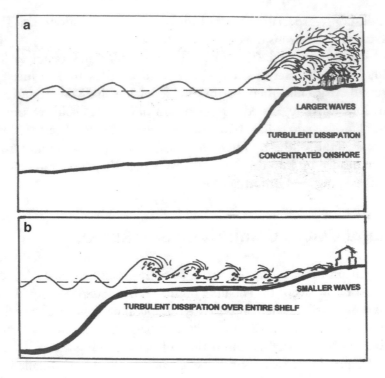

Fig. 4.10 (a) Tsunami waves moving over deep water carrying much of their energy to shore. (b) waves over a shallow shelf loose much energy producing smaller waves ashore

In contrast, the opposite effect occurs where tsunamis (and large storm waves) move at high speed over the deep coastal water (i.e. absent a wide shelf) carry much of its energy close to shore (see Fig. 4.11). Upon meeting abrupt shoaling, they concentrate their energy by suddenly growing to great heights. Extreme inland flooding ensues over low-lying coasts. This was the case in Hawaii in the 1960 tsunami, as it moved in from deep water, ran up the steep sided volcano, developed wave heights of 7–10 m and flooded the city of Hilo.

Hurricane waves coming ashore can have a more complicated scenario. They are often riding upon a "storm surge" whose added water height (typically ~2–5 m above MSL) is pushed like a tide onto shore by approaching winds. The hurricane waves thus are added to the storm surge, which increased both run-up height and distance. Also, note that the wind wave run-ups are still limited by the wave periods which generally don't exceed ~30 s. Thus the oncoming waves averaging ~10 m/s would only advance ~150 m during the half period of the waves.

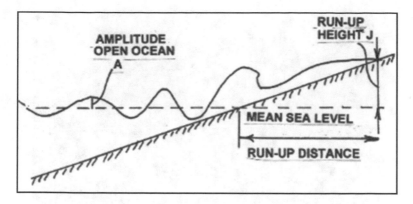

Fig. 4.11 Wave coming ashore defining run-up distance and height

For tsunamis it's another story. With their huge wavelengths (often several hundreds meters long, even at the coasts), and ultra-long periods (generally ~15–30 min). They can have run-up distances of several kilometers over the half period of 7–15 min. As the first large wave crashes ashore, the sea behind eerily flows like a wide river for half the wave period, then recedes for the second half. However, often only part of this water recedes, causing the water to build up onshore as the cycles inexorably repeat—oscillating back and forth. This suggests a rapid, deliberate tide-like motion wherein the term "tidal wave" (a misnomer) probably originated.

The Cortes Bank

There are places in the Earth's oceans where unique bottom topography along with refraction cause storm waves to metamorphose into incredible monsters that surpass the heights of most tsunamis creating a "run-up" running amuck. One example is over the Cortes Bank off the coast of Southern California where the bottom abruptly rises from a 1400 m depth to within a few meters of the surface (Dixon 2008). Spawned by winter storms in the North Pacific, the Bank receives long swells, propagating eastward with 200–300 m wavelengths and heights of 5–8 m. Nearing the Bank, the swells encounter the rapidly shoaling water where refraction focuses them onto the steep bank. The result is an explosive upsurge over the reef, producing some of the world's highest breaking waves—upwards of 30 m (see Fig. 4.12). As evident in the photo, daring (or crazy) surfers have discovered the Bank: "Surf's (really) up!"

Fig. 4.12 The giant waves breaking over Cortes Bank, California

We now briefly review examples of the most extreme tsunamis on record. For comparison, this will then be followed by models that depict hypothetical tsunamis—each with much greater energies then any recorded events, yet much dwarfed by our final portrayal of the monster tsunami borne out of the Chicxulub impact.

The Indonesian Earthquake Tsunami: 2004

As we have discussed, the Indonesian Earthquake of 2004 spawned one of the most powerful tsunamis ever recorded. As noted in Fig. 4.8a–c, the tsunami waves moved directly eastward from the earthquake epicenter to the tourist beaches of Sumatra and Thailand. Within 20–30 min after the quake, waves inundated the Khao Lak beach in southern Thailand as shown in the photos taken by a Canadian tourist (Pilet 2007), which record the devastating waves (see Fig. 4.13a–c).[19]

The first tsunami crests (see Fig. 4.13a) were seen building 3–4 km offshore. The local beach water level appeared to recede seaward, signaling that the

---

[19] The photos were recovered by Pilet (2005) from a digital camera memory card found on the beach, a month after the tsunami. The photos were taken by John and Jackie Knill, Canadian vacationers, as the waves proceeded onto the beach. Tragically they, along with some 5000 others, lost their lives along the Thai beaches.

Fig. 4.13 The Indonesia Tsunami waves breaking over the Khao Lac beach in southern Thailand. (**a**) Waves building up and breaking over the barrier reef several km offshore. (**b**) Water near shore ran out as the trough approached. With the horizon filled with huge turbulent breakers, force of waves coming ashore smashes upon the reefs with heights over 12–14 m. (**c**) The huge crashing waves finally reach the beach at heights of 10–20 m and commence to flood the coast (permission by Knill family)

tsunami's wave trough arrived first.[20] (This likely encouraged bathers, unaware of the danger, to stroll seaward to explore the exposed reefs—instead of immediately seeking high ground.)[21]

Minutes later (see Fig. 4.13b) the leading crest broke over the outer barrier reef, sending surf 8–10 m upward while the water on the beach continued to draw outward. The full fury of the towering wave crest poised prior to crashing over the moored fishing boats while a lone bather looks on. Finally, the waves with estimated heights of 10–12 m (see Fig. 4.13c) reached the beach blocking out the entire horizon with a powerful seething mass of turbulence. The waves then advanced inexorably hundreds of meters inland, sweeping up cars, beach houses, trees and hotels in a giant racing tide. After 10–15 min, having advanced inland 1–3 km the watery avalanche slowed, stopped, and then reversed, carrying much of its debris seaward. This surge producing cycle that repeated, oscillating on and offshore currents with 20–25 min periods lasting for several cycles.

The Tsunami at Unimac Island, Aleutians

One of the most extreme run-up heights was experienced by a tsunami generated by a strong undersea earthquake on 1/Apr/46, centered 125 km southeast of Unimac Island in the Aleutian Island chain in the North Pacific (Shepherd et al. 1950). Waves arriving from deep water probably with maximum energy, encountered an abrupt shoaling running up to a steep cliff on the island that sealed the fate of the Scotch Cap Lighthouse (see Fig. 4.14, left panel).

Two or three major waves came ashore at night, channeling up the steep slope and demolishing this cement structure at 27 m above sea level and drowning all Coast Guard personnel (Fig. 4.14, right panel). The waves finally crested above the lighthouse, bursting over the top of the cliff and sweeping away a large antenna, with an estimated run-up height over 35 m.

---

[20] Note the first part of a tsunami wave to arrive on shore can be either the trough or the crest.

[21] This scenario is historically the most deadly experienced by bathers at a tsunami event. And is why school children around the Pacific coasts are taught that if beach water starts to move out to sea—or if you stop hearing the surf—run—from the shore to higher ground.

Fig. 4.14 Results of Tsunami moving in from deep water onto Unimak Island, cresting up to a run-up of over 35 m demolishing the concrete Scotch Cap Lighthouse and upper antenna structures (US Coast Guard Photos)

The Great East Japan Earthquake Tsunami: 2011

Finally, we consider a strong tsunami set off by one of the most powerful earthquakes of this century. On Friday 11/Mar/2011, an undersea thrust, centered 70 km off the northeast coast of Japan, produced an extreme earthquake of magnitude 9. Figure 4.15 illustrates the peak of the harbor flooding at Miyako (JiJi Press, Toru Yamananka, (2011).

The first crest appeared as a 10–12 m super high tide, moving over the entire harbor, overtopping the seawall onto the city. In other locations several coastal villages were completely swept away with wave run-ups over 25–40 m high inundating low-lying coasts and in some areas moving 10–12 km inland.

The above examples document tsunami waves produced by the most powerful earthquakes. However, the largest waves produced seem to rarely exceed 15–25 m in height and run-up heights rarely exhibit run-up distances of 5–15 km. It is suggested that tsunamis created by giant earth slides, as well as asteroid impacts, will produce far greater wave events. In order to explore the question of how big tsunamis can form, we consider two prediction models, each of which could conceivably generate waves larger than any recorded to rampage coasts of the North Atlantic Ocean.

Fig. 4.15 The initial tsunami wave flooding of Miyako Harbor, Japan from the massive Earthquake of 11/Mar/11 (JiJi Press, Toru Yamananka)

The La Palma Volcano Collapse Tsunami: A Model

As we have said previously, next to Earthquakes, the most common cause of tsunamis is the lateral collapse of volcanic island calderas or of built-up lava flows, which spontaneously slide into the ocean. Evidence of such events has been found around the bases of volcanic islands worldwide.[22]

---

[22] The Hawaiian Islands exhibit the most prominent examples of caldera collapses with massive slide debris piled up on the sea floor at the base of the steep sided volcanic slopes.

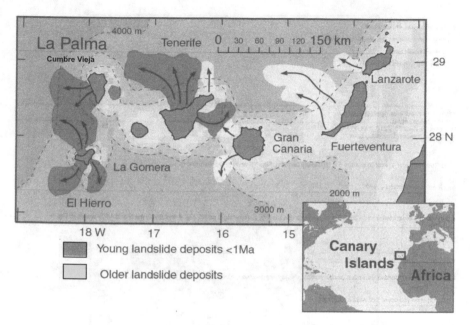

Fig. 4.16 La Palma Island in the Canary Islands off the West African coast showing the many landslide deposits associated with volcanic eruptions. The active volcano Combre Vieja is located on the NW flank of La Palma (by Ward—drawing)

Of interest is the volcano Combre Vieja, on the Island of La Palma in the Canary Islands off the African west coast (see Fig. 4.16). Around this island the sea bottom is littered with massive volcanic debris fields, extending outward some 50–100 km. These are accumulations of volcanic material that was produced during periodic eruptions, when its caldera collapsed and slid down the steep 15–20° slopes into the ocean. Clearly such events set a perfect stage for powerful tsunami generation.

Poised on the northwest tip of La Palma is the 2000 m Cumbre Vieja, the most active volcano in the Canaries group, having had seven eruptions over the past 500 years. During the last eruption, in 1949, a ~4 km long fault opened up on the west flank of the caldera. Part of this ridge slipped 2–3 m downward toward the open Atlantic Ocean—and stopped. Surveys suggest the fracture was of a wedge shaped slab ~20 km long which extends downward to 1–3 km beneath the sea surface. Estimates of its volume range up to ~500 km³, or some 380 Rocks of Gibraltar.

Of major concern, is that the volcano could be a ticking time bomb. To wit, a future eruption could release the monster slab into the sea, producing a major tsunami much of whose energy would be focused westward, directly

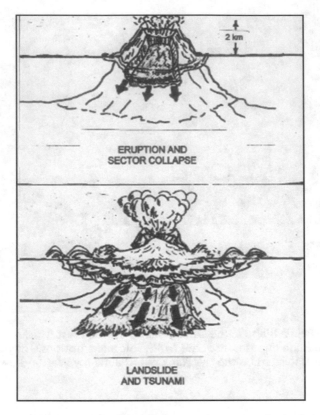

Fig. 4.17 Schematic of the volcanic eruption and release of a sector of the volcanic caldera. As it slides into the ocean (*upper panel*) the tsunami is produced forming a giant water dome (*lower panel*) whose vertical oscillations produce the tsunami waves moving radially outward redrawn from Ward and Day

threatening the east coast cities of North and South America (including the low lying Cape Canaveral, FL with its NASA Space Center). The question was posed, by Dr. Steven Ward of the University of California and Dr. Simon Day of University College, London, who hypothesized that a future eruption of Cumbre Vieja could loosen the ~500 km³ block ~25 km long, ~15 km wide and 1400 m thick (Ward and Day 2001). The huge mass (see Fig. 4.17), could then slide down the steep western slope into the ocean, reaching ~60 km out over a flat sea floor in a water depth of approaching 4000 m.

Their model, employing classical wave theory, predicted that within 2 min of the collapse, the enormous mass sliding into the sea and forming a dome of water ~900 m high. Within minutes, the collapsing dome would begin to oscillate giving birth to a train of giant tsunami waves 300–600 m high moving radially outward from the island. After 5.5 min the sliding block, moving at 100 m/s (225 mi/hr) would soon be overtaken by the accelerating tsunami waves which, when moving into increasing depths of >2000 m would exceed 150 m/s (340 mi/hr). By 10 min, the western moving wave heights would drop down to ~250 m while wavelengths increased.

Within the next 45 min refracting waves, wrapped completely around La Palma, impacting the eastern-most Canary Islands, adding a hundred or so meters to their heights as they swept ashore. Within an hour (Fig. 4.18), the 50–100 m high breakers hammered the West African coast, spelling disaster for hundreds of coastal villages.

Three hours later, the center of the 6000 km arc of westward heading waves would have reached halfway across Atlantic Ocean, where wave heights were now reduced to 20–30 m. After 6 hrs (see Fig. 4.18), the southern segment of the giant arc of wave energy started to inundate South American coasts where the waves slowed hitting shallow depths, grew by tens of meters and flooded beaches and coastal villages in Brazil and Venezuela. Meanwhile, to the far north, 10–15 m high waves made landfall at Newfoundland. And, by 9 hrs out, tsunami waves would have battered the entire US east coast with wave heights of 10–20 m. By this time, in Florida, waves moving over its narrow shelf grew to 20–25 m. The central and southeast coast of the Florida peninsula (see Fig. 4.19), because of its low altitude of 3–5 m, would be especially vulnerable to the oncoming breakers. The 2–2.5 min period waves probably allowed run-up distances of 2–3 km along the coasts—a disastrous deluge. The waves of 10–30 m heights hit Florida beaches, bordering coasts having only 2–5 m altitudes, severely flooding coasts north of Palm Beach. South of Palm Beach to beyond Miami, tsunami waves were much smaller due to the sheltering of the Bahamas Island and the vast banks.

Clearly, a Cumbre Viejo collapse is a prognosis for grim disaster along the coasts of both sides of the Atlantic Ocean. In part because of an extremely short warning time: the West Saharan shore would have a mere ~1 hr before the tsunami struck, while cities on the East Coast of North and South America would have only 10–12 hrs. The run-ups and flooding would be much greater than from the worst of hurricanes or the most powerful earthquake tsunamis. Moreover, unlike hurricane events, tsunami flooding could occur along the entire coast of North and South American—within 1–2 hrs! Results of this model serve as a stern warning to municipalities on the Atlantic American East coasts to consider developing evacuation plans to deal with a direct shot from a westward bound La Palma tsunami express.

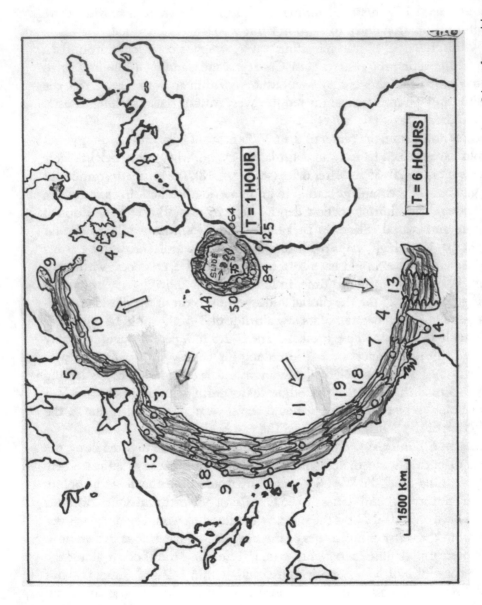

Fig. 4.18 The spreading Combre Vieja wave packet model positions in the Atlantic Ocean at 1 an 6 hrs after caldera collapse. The wiggly lines suggest the wave packets consisting of ~10–20 largest wave heights. Mean wave heights (m) positions indicated by white circles (redrawn from Ward and Day)

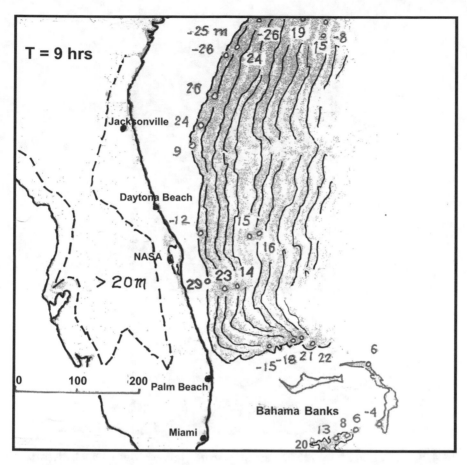

Fig. 4.19 Tsunami waves approaching the Florida east coast 9 hrs after caldera collapse. Coastal shoaling causes the waves to slow and build to 20-15 m. The coastal plain east of the 20 m isobath is flat and only a few meters above sea level. Coastal cities would suffer severe flooding, except for south of Palm Beach where the Bahama Banks shield the coast). (Suggested by Ward and Day (2001)

The 1950DA Ocean Impact Tsunami: A Model

As we discussed in Chap. 1, the asteroid 1950DA at 1.1 km diameter, is scheduled to pass (at least) uncomfortably near the Earth in the year 2880. Scientists, Drs. Steven Ward and Eric Asphaug, from the University of California, using wave theory similar to that used to predict the Cumbre Viejo tsunami, constructed a wave model generated from the hypothetical impact of 1950DA into North Atlantic Ocean, 600 km east of the United States east coast in ~5000 m water depth (Fig. 4.20).

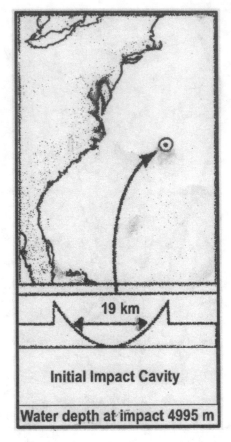

Fig. 4.20 Location of the 1950DA asteroid ocean impact over ~5000 m depth. The lower profile suggests the water crater height of ~5000 m (Ward and Asphaug 2003)

They assumed the object would make a vertical impact at ~17.8 km/s, have a density of 3000 kg/m³; and would have a KE of ~2.5 × 10²⁰ J. (Compare this with others given in Table 3.1.) The impact would blast a water crater of ~19 km in diameter and ~5 km deep. Their model doesn't make the asteroid "bottom out," which complicates the energy calculations, so they assume the initial wave height equal to the bottom depth ~5000 m.[23] The initial largest wavelength is predicted ~ equal to the crater diameter and at the long wave speed for a depth of 5000 m, translates to a period of ~2 min.

This model simulates tsunami waves as they radiate over the Atlantic from ½ to 9½ h after impact. Figure 4.21 (based on the Ward and Asphaug paper 2003)

---

[23] As we will see further, this may not be the rule in the case of 10 km Chicxulub impacting the ancient Gulf of Mexico with variable depths of 500–1500 m. We suggest that the explosion blows water outward creating a pile-up crater with a rim height much greater than the depth—the Edgerton Effect.

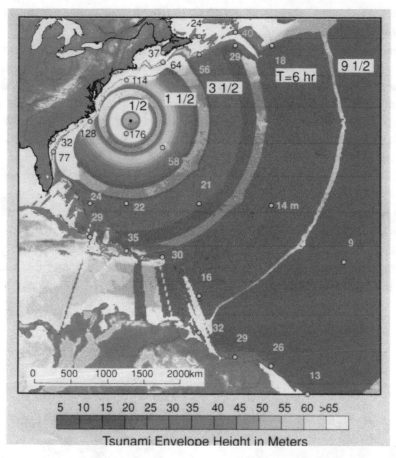

Fig. 4.21 Schematic of the time generation of the spread of the tsunami over the Atlantic Ocean. Note the waves decrease in height as they spread eastward whereas the westward wave greatly increase as they approach the coast. Wave heights of 100–150 m move over beaches from Halifax to Florida (Ward and Asphaug 2003)

shows time sequences of the tsunami waves as they spread out and approach the Atlantic coast. Estimated wave heights are located at the white dots.

By one half hour after impact, the initial 5000 m wave height above the crater has decreased to ~175 m. After 1 hr, the westward moving waves start to cross the continental shelf, slowing down and increasing in heights. Then (as Ward and Asphaug put it) the giant arc of concentric waves would change into "flat tire" shapes. At 1–2 hrs out, the waves continued to grow in height as they approached shallower water, reaching to ~120 m on the beaches from Cape Cod to Cape Hatteras—an absolute disaster! Their run-up heights were predicted to be in the range of ~70–140 m (some three times the size of their offshore heights).

Again, as with the La Palma waves, the Florida peninsula's low altitude and flat terrain makes it a sitting duck—particularly vulnerable to the oncoming breakers. With repeated poundings of 20–50 consecutive wave crests, even a 2 min wave period would allow water to accumulate, surge on shore up to perhaps even 2–5 km, sufficient to severely flood, the coastal cities from Palm Beach to Jacksonville.

Within 3–9 hrs, the tsunami, with 30–35 m high waves would reach the Windward Islands of Cuba, Haiti, and Puerto Rico in the northern the Caribbean. In the mid-Atlantic Ocean, eastward moving waves would have decreased to ~9 m while the southerly-directed waves build up on the shallows of South American coasts to heights of 25–30 m.

The Eastward bound Atlantic tsunami waves, after about 8–9 hrs, would have reached the European coasts (see Fig. 4.22, shaded areas) from northern Spain down to the North African coasts. To the north, the waves slowed

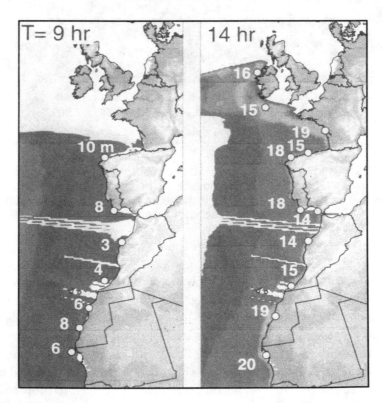

Fig. 4.22 The eastward heading tsunami at 9 and 14 hr after impact. North of Spain the shelf slowed the waves (*white*) approaching the British Isles while the heights increased as waves crashed upon European and North African coasts (Ward and Asphaug 2003)

significantly as they reached continental shelf around the British Isles. By 14 hrs, waves having grown to 15–20 m caused major inundations of coastal towns and cities from Scotland clear down to the west coast of North Africa.

The primary damage to the Earth's environment from the 1950DA's ocean impact was associated with the monster tsunami which would have drowned coastlines on both sides of the Atlantic. The large ocean volume would largely absorb the heat from the explosion. Were the modeled impact of 1950DA to occur in today's world, many cities would be destroyed. Moreover, without accurate prediction (unlikely) of the impact location and advanced warning, millions of lives would be lost.

Ward et al. emphasized that the modeled 1950DA impact should serve as an alert to keep a watchful eye on NEO objects that have even the slightest potential for Earth impact. And, that although the human race has yet to experience such a disaster, some 200 such 1950DA-sized asteroids have likely struck Earth since the time of Chicxulub. We now turn our attention to the story of the tsunami likely generated from the ultimate asteroid ocean impact: The 10 km Chicxulub crashing into the ancient Gulf of Mexico.

The Chicxulub Impact Tsunami

Our description of the Chicxulub tsunami and its effects on coastal environments is based on ancient historical geological evidence with the help of theoretical models, illustrated in this chapter. Direct evidence of these great waves is suggested at outcrop sites where ancient coastal ocean sediments appear deposited at various heights above the ancient sea level around the Gulf and Caribbean (e.g., in Haiti). Moreover, these sediments were found in the same or adjacent strata as the K-Pg boundary material, thus linking the tsunami with the impact event.

The Prologue to this book, characterized the onshore arrival of a series of giant waves that inundated the low-lying coast below our mythological vantage point in western Cuba. Since, this giant splash generated tsunamis far more powerful than any known earthquakes, or volcanic eruptions, we ask how well can we extrapolate the estimated effects of, e.g. 1950DA, to portray the scenario of the Chicxulub tsunami?

As we have seen, the size and power of the tsunami generated from an ocean asteroid impact depends upon a variety of variables such as: (1) the size, (2) object speed (i.e., its kinetic energy) and (3) its angle of attack. Another important variable could be the object's size compared to the water depth. This seems especially critical in comparing Chicxulub with 1950DA tsunami.

For the 1950DA hypothetical impact modelled previously, the asteroid's diameter was 1/5th of the ocean depth so the splash was more like the stone hitting the pond. Most of its KE was imparted to the water, causing it to compress, vaporize then explode, throwing liquid water, vapor, and asteroid fragments into the atmosphere. The turbulent splash, formed at the sea surface. And, as it quickly dissipated, the ensuing collapse left a vertically oscillating mound of water, producing symmetric long waves that spread concentrically outward (see Fig. 4.21).

For Chicxulub, the scenario was wildly different. With its diameter at ~ten times the water depth, the impact explosion immediately "bottomed out." The relatively thin water layer had little consequence upon the impact dynamics with most of the asteroid's KE absorbed into the explosion and excavation of the huge sea floor crater.

The explosion on the initially shallow water had to have produced enormous waves, but the scenario was more complex than just a scaled up version of a proverbial stone splashing into a pond. Noting that the water depth over the impact site may have ranged from 500 to 1000 m, the best analogy of this impact is more likely that of a giant rock smashing into a lake—and blowing most of the lake away!

It was suggested in Chap. 3 that during the formation of the "transient cavity crater", the Gulf water, not instantly vaporized at impact, was blown outward to (and beyond) the 180–200 km crater edges creating a giant water annulus. The edifice, some 5–10 km high, then collapsed triggering the tsunami. What would this have looked like?

The Chicxulub Tsunami's "Edgerton Effect"

The Chicxulub tsunami generation might be visualized as analogous (but on an extremely smaller scale) to a sequence of famous high-speed photographs of the impact of a milk drop falling on a wet table taken by Professor Harold Edgerton of the Massachusetts Institute of Technology (see Fig. 4.23a–f).[24]

At impact (Fig. 4.23a), the drop flattens with its outer edge deflecting upward (b) forming a crown-like shape (c). The remainder of the liquid inside the crown accelerates radially outward. At (d) the liquid, which was scoured out from the impact zone, expands and piles up forming the annular "crown pattern." Finally, as the wave slows down, its crest breaks into smaller pieces (e), which continue to fly outward (f). The height of the outer ring is roughly the diameter of the original droplet. Note, also, that instead of a turbulent

---

[24] "Doc" Edgerton's inventions and activities were endless: he developed the stroboscopic flash seen on beacons worldwide, photographed the atomic bomb tests at ultra-high speeds, helped develop the "side-scan" sonar, photographed the sunken civil war ship "Monitor," searched for the Loch Ness Monster, and built undersea cameras for Jacques Cousteau—who affectionately called him "Papa Flash." Doc was truly a "man for all seasons."

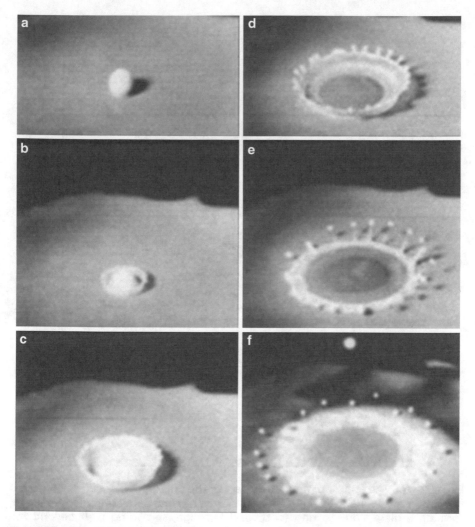

Fig. 4.23 The Edgerton milk drop photograph sequence suggesting how the Chicxulub impact produced the horizontal momentum which formed the great annular ring whose collapse generated the tsunami (Photo courtesy of the Edgerton Center, 2015 with permission of the MITMuseum, 2016)

splash, the liquid is propelled bodily outward some ten object-diameters before forming the near perfect annular ring with breaking crests.[25]

The resulting phenomenon that we call the "Edgerton Effect" suggests how, from the Chicxulub explosion, water was propelled away from the impact area and formed an enormous annular ring. The sequence of the crater formation

---

[25] It is cautioned that surface tension may suppress the splash effects that would normally accompany an ocean impact. Moreover, the analogy stops once the outer annulus is formed, since the rigid tabletop prevents the inner formation of the "bottom" crater.

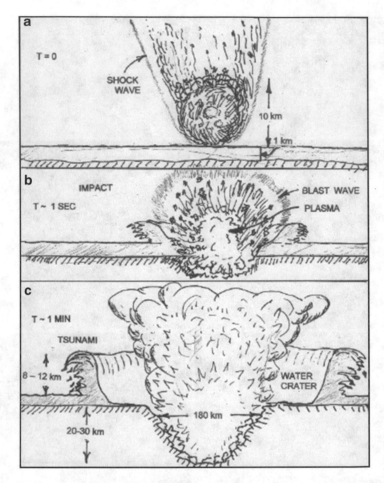

Fig. 4.24 (a) The approach and the Impact of the Chicxulub object (b). Suggested formation of the mega—Edgerton Effect when the blast pressure piled up water, forming a giant annulus ring (c) several km high, spreading well beyond the bottom Earth crater

is suggested in Fig. 4.24a and b where the explosion skims off the water layer forming the growing water annulus Fig. 4.24c.

The final annulus would have taken on huge dimensions because of the vast amount of seawater forced outward. The final annulus whose outside diameter extended 250 km with a height of 8-12 km. The collapse and subsequent oscillations of this liquid colossus, initiated the Chicxulub tsunami.

The monstrous waves, as they moved over the 500–1000 m water depths, immediately transformed into classic long waves conforming to hydrodynamic laws and obeying the "Common Sense Rule," i.e., stable wave heights cannot exceed the water depth: this would cause their troughs to be on land—physically impossible. Thus, as the waves moved away from the oscillating water crater annulus, both their speeds and heights quickly adjusted to the changing water depths. Initially, the tsunami wavelengths were roughly that of the water crater diameter (~200 km) and the periods of oscillation (~30 min)—about the time required for the long wave to traverse the crater. The oscillating Chicxulub tsunami machine was now giving birth to the huge waves racing outward in all directions.

Chicxulub's Spreading Waves and their Global Reach

How Chicxulub's giant tsunami was to spread over the world's oceans and flood the continents was governed in part by two parameters: (1) its inherent energy associated with the initial wave heights and (2) the morphology, i.e., the size, shape and depths of the oceans over which the waves spread. Since the Chicxulub impacted near the center of the ancient Gulf (see Fig. 4.1), the tsunami waves spread out radially, onto the nearby coasts except for the those heading eastward to the deeper Atlantic Ocean and northern Caribbean (see Fig. 4.25).

Within ~2 hrs after impact the first waves reached the mountainous coast south of the ancient Central America and the western tip of Cuba as depicted in our scenario in the Prologue.[26] Wherever the topography rose up sharply from deeper water (Fig. 4.10) the tsunami would have slowed down, quickly adding hundreds of meters to the wave heights as they broke over the beaches. However, the run-up would have extended only a few kilometers, having been impeded by foothills and steep mountain slopes.

Within another hour or so, the tsunami waves heading west and north still probably exceeded 300–500 m heights, reached the flatter coastlines. The northward running waves invaded the great inland sea in the middle of North America (discussed in Chap. 3.) These waves, moving over even shallower water with crests of ~100 m, flooded inland over the Great Plains even as far north as Colorado and Nebraska. Entering the shallow depths of ~50–100 m, these waves were reduced to 20–30 m breakers as they crashed onto western and eastern beaches bordering the foothills of the Rocky and Appalachian Mountains.

Around the ancient Gulf area the pounding of the towering breakers upon the ancient shorelines caused violent land erosion and turbulent mixing of bottom sediments, drowning of thousands of kilometers of coastline. Intense

---

[26] The Cretaceous Cuba and the Windward Island chain were likely narrow strips bordering the northern Caribbean being pushed up by the spreading Atlantic plate.

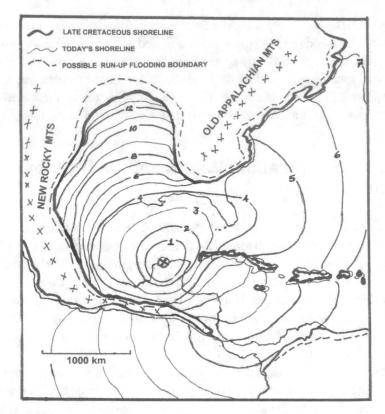

Fig. 4.25 The wave fronts of the tsunami radiating out from the impact point at roughly hourly intervals into the ancient Gulf and thence, to the Caribbean and the Atlantic Ocean

opaqueness of the mud and sediment-laden waters together added to the now blackening skies further enhanced the rapid shut down of photosynthesis in the ancient Gulf waters and beyond.

Southeastward-directed waves headed for the northern Caribbean, their speeds picking up to 200–250 m/s (450–570 mph) over the deeper waters as crest heights fell to 200–400 m. As the waves moved ashore they added 100–200 or so meters to their heights, completely inundating hundreds of low-lying Caribbean islands. Such islands, formed, as by-products of reef building had altitudes of ~1–4 m, and were completely submerged for many hours if not days.

Since Florida and the Great Bahamas Bank did not yet exist, major eastward moving Chicxulub tsunami waves surged uninhibited out of the Eastern Gulf, with a clear unobstructed shot across the north and mid—Atlantic Ocean (see Fig. 4.26). The shaded areas suggest where there occurred maximum coastal run-up, drowning and erosion.

As the waves spread over the Atlantic's deep abyssal plains, at 4–5 thousand meter depths, their speeds increased to ~200–250 m/s (400–500 knots) while

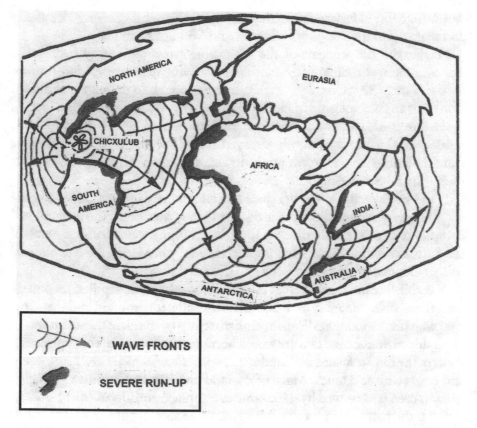

Fig. 4.26 The global spreading of the tsunami waves. Most of the energy was directed eastward, into the Atlantic Ocean, then spread out to the South Atlantic and through the Indian Ocean. The shaded areas depict severely eroded and flooded coasts

their heights likely decreased to below 200 m (compare this to the earthquake tsunami waves in the open ocean normally having ~0.5–1 m heights).

Crossing the Atlantic Ocean within 5–7 hrs, Chicxulub waves hit the European and African coasts, refracting over the coastal shelves. The shoaling waves slowed down as their heights built up, forming towering 200–300 m high breakers that inundated the coasts from Scotland south to West Africa, where waves often invaded tens of kilometers inland.

Within 10–15 hrs, the wave fronts refracted around and onto the coasts of South America and Africa. Then rounding South Africa, they raced eastward through the ancient Indian Ocean, inundating those coasts and islands with still 100 m high breakers.

The great southern arm of the Chicxulub tsunami whose wave heights, now were reduced to below 100 m, followed the ocean ridges, and moved eastward across the north and south sides of the ancient island-continents of India and

Australia. Most of its wave refracting energy dissipated as crashing turbulence and foam upon the shores of Indonesia and Western Australia.

Finally, the last segment of the Chicxulub tsunami's energy had spread out over the vast Pacific. After another 10–12 hrs, the waves, reduced probably to a mere 30–50 m high, were still large enough to drown thousands of low-lying ancient volcanic Islands in the Western and Central Pacific Ocean. Spreading further eastward, the tsunami experienced its last hurrah, as its modest 10–15 m waves slowly decayed, soon becoming indistinguishable from the ambient ocean storm swells that crashed upon the west coast beaches of North and South America.

It had been barely 24–36 hrs since the mighty fiery impact. Following the traumatic explosion and the resulting continental firestorms, the tsunami had completed its own devastation. Its towering crashing waves had left much of the Earth's coastal land areas and islands, especially bordering the Atlantic and Indian Oceans, in crushed, eroded and drowned ruins.

A sizable fraction of the Earth's coastal biota—the birds, reptiles, mammals, insects and plant life must have been terribly battered, drowned and then covered with black muddy acidic sediments stirred up by the giant turbulent waves. Soon this environmental disaster would become amplified with the commencement of the impact fallout and muddy rains and snows which filled lakes, rivers, and coasts with acid muck. Much of the land on Earth had, within a geological instant, been transformed by the cosmic event into a semi global "silent spring."

With the passage of months and years, more subtle, but inexorable effects from the poisoned and blackened atmosphere took hold as a long-term inevitable global climate upheavals commenced.

References

Dixon C (2008) Surfers defy giant waves awakened by storm. (Surfing) New York Times, 26 Jan 2008 C16

Pilet C (2007) The couple on the beach. Guideposts 61(11):57–58

Shepherd F et al (1950) The tsunami of April 1, 1946. Bull Scripps Inst Oceanogr 5(6):391–528

The Edgerton Center (2015) MIT. http://ocw.mit.edu/courses/edgerton-center/. Accessed 15 Dec 2015

Titov V et al (2005) The global reach of the 26 December 2004 Sumatra tsunami. Science 309:2045–2048

Ward S, Asphaug E (2003) Asteroid impact tsunami of March 2880. Geophys J Int 153:F6–F10

Ward S, Day S (2001) Cumbre Vieja volcano: potential collapse and tsunami at La Palma, Canary Islands. Geophys Res Lett 28(17):3397–3400. doi:10.1029/2001GL013110

5

Long Term Global Effects

Abstract Chapter 3 described how the sudden impact of Chicxulub upon the ancient Gulf of Mexico caused a rapid-fire sequence of disturbances, some occurring within minutes, others within hours or even days. In this chapter we summarize the longer-term effects of Chicxulub's devastating impact on the oceans, atmosphere and land areas. This includes changes in the oceanography of the area from the mixing of hot toxic seawater produced in the impact crater. Also discussed is the alteration of global weather and climates due to the massive influx of sooty particles, gas ejecta and boiling crater vapors into the atmosphere. First is the global cooling associated with the blocking out of the sunlight, then a rebounding greenhouse warming smothering the globe for years? We then note the effects of Chicxulub on the fate of the dinosaurs the other animal populations. Finally we consider the survival of the small mammals and how their presence may have affected the evolutionary path leading the advent of ancient and modern mankind.

Oceanographic Chaos: Creation of the Hot Toxic Sea

We discussed in Chap. 3 how the Chicxulub explosion initially excavated a skillet shaped hole in the bottom of the ancient Gulf of Mexico, throwing much of its ejecta into the atmosphere. The immense transient crater had an area

© Springer International Publishing Switzerland 2017
D. Shonting, C. Ezrailson, *Chicxulub: The Impact and Tsunami*,
Springer Praxis Books, DOI 10.1007/978-3-319-39487-9_5

of more than 20,000 km² (about that of New Jersey) and had a base depth reaching 20–30 km. Soon this transient cavity collapsed reducing the depth to 10–15 km, still far deeper than say, the Grand Canyon (see Fig. 3.12). As we have noted, the impact blast reached plasma temperatures of thousands of degrees, transforming the cavity into a volcano-like edifice. Interior rock debris, not thrown skyward, liquefied, and pored into the crater base forming a single "griddle cake" melt sheet (Fig. 3.11d) which became a gigantic blast furnace-like heat source.[1]

Had the crater been created on land, its fiery contents would have been isolated from the surrounding area while its mass gradually cooled. However instead, the impact site, located in the ancient Gulf, produced far different attributes of the post impact environment. Dramatic changes must have occurred in the atmosphere and especially in the oceanographic conditions in the surrounding sea environment. The key variable associated with the imposed changes in oceanic properties was in the parameter of "stability".

Ocean Stability Effects

Under normal conditions the oceans form a layered structure where its density (mass per unit volume), which is determined mostly by its temperature, generally increases with depth. This is shown in Fig. 5.1a where the strong temperature gradient or "thermocline" occurs in the upper layer (i.e., where the temperature profile decreases rapidly with depth), which causes the density to increase rapidly with depth.

Within the upper layer the warmer, hence lighter, water layer floats atop of cooler, heavier water below. Thus, in this layer at a given depth a water "parcel" is both denser than the water above but less dense than the water below and thus resists moving up or down, i.e., vertical mixing is inhibited, resulting in what oceanographers term, a "stable" condition.

However, when periodic seasonal cooling or extreme weather causes the surface water to become colder (denser) than the water below (see Fig. 5.1b), the upper water column becomes top heavy or "unstable" causing it to sink. This effect is accelerated by strong winds at the sea surface, which generate turbulence, adding to the downward transfer of surface water. This provides the mechanism whereby deeper waters are periodically replenished with sur-

---

[1] Note that the "geothermal gradient" (the natural heat increasing with depth in the Earth's crust) is about 20 °C/km (due mostly to the Earth's radioactive heating) so the deep exposed rock may have exceeded 300 °C. This allowed the crater to sustain its thermal heating long after the melt sheet cooled down.

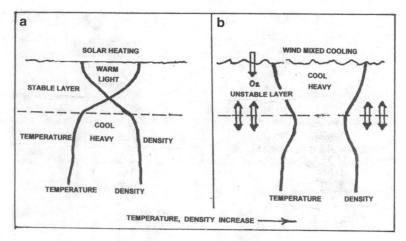

Fig. 5.1 (a) Typical ocean temperature profile showing a steep gradient through the upper layer. This decrease in temperature, or thermocline, produces a strong increase in density with depth where vertical mixing into the deeper layer is inhibited. This creates the "stable" layer of warmer—lighter water which floats upon the deeper layer where both the temperature and density change more slowly with depth. (b) With strong cooling from the atmosphere and heavy wind stirring, the surface layer can temporarily become cooler, well mixed, and heavier than the layer below. These conditions are unstable and soon the upper layer collapses and extreme vertical mixing occurs. It is such intermittent mixing events which allow oxygen and nutrients to be transferred from the sea surface to the deeper ocean layers

face oxygen and nutrients. In the world's oceans there is a sustained air-sea interaction so that the upper ocean layers receive on a seasonal basis (on average) enough oxygen and nutrients to sustain sea life.

There are cases where extreme strong local heating of the surface layer occurs, causing it to become much more buoyant. This layer can act as a lid, suppressing the downward mixing of oxygenated or mineral-rich surface water. If such "hyper" stable conditions persist, it can be catastrophic to sea life, since the bio uptake in the deeper layers seriously depletes oxygen and nutrients.

Such extreme "super" stable conditions were likely generated around the hot Chicxulub crater. Over the days, even weeks, as the crater filled with Gulf water its contact with the deep superheated melt sheet produced huge steam explosions and heating of the water, creating a enormous convection system (Fig. 5.2). The sequence begins with water, pouring into the bottom of the crater, which becomes superheated as it flows over the hot melt sheet and shattered breccia. It also mixes with deep rocks, which we recall contained

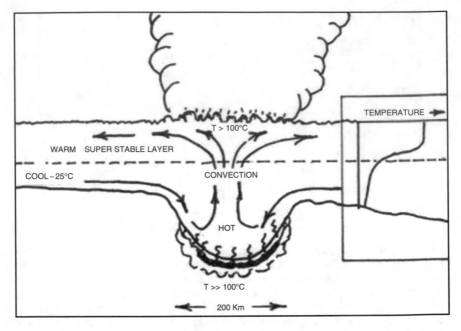

Fig. 5.2 The huge convection system driven by the Gulf seawater being heated by the melt sheet and toxic debris at the bottom of the impact crater. The boiling water mixed with the sulfur rich minerals rises by buoyancy to the surface. A part boils into the team cloud and part spreads out forming a hot stable surface layer. *Insert*: The temperature profile away from the crater delineates the hot super stable surface layer. Its super stability inhibited vertical transfer of surface oxygen and nutrients to the deeper depths for months if not years

an unusual high content of sulfur minerals. This process would produce an especially acidic and toxic fluid to admix with the heated seawater. Meanwhile the heating increases the buoyancy of the deep water causing it to bubble up rapidly to the surface, carrying with it both heat and the dissolved toxic chemistry. As the hot toxic water convects upward from the bottom it is replaced by the adjacent cooler water coming in from the side. This new cooler water is then heated up and continues its cycle to the surface and then spreading outward.

Still in a boiling state, some of the water escapes into the atmosphere, but most forms a hot super stable layer which, with the help of surface winds, spreads horizontally over the Gulf, the adjacent Caribbean and possibly over the Eastern North Atlantic waters. The hot toxic layer formed a super buoyant stable layer, suggested by the temperature profile (Fig. 5.2–insert), perhaps several hundred meters thick and after time extended hundreds of kilometers out from the crater zone which isolated the deeper water from the surface layer.

Such was a perfect storm of disasters in the waters above the boiling crater system: Poisonous hot waters spreading over the surface layers which overlay a deeper anaerobic layer provided no safe place for the poor fish and animal life to go. To complete the catastrophe, months of acid rains pelted the sea surface while sediment leaden coastal runoff completed the recipe for both poisoning and suffocation of life around the Gulf and probably far beyond.

The Long Dark "Nuclear" Winter

Within hours of the impact explosion, the huge ejecta plume had reached the upper atmosphere where the winds of the Jetstream soon spread the ash-laden cloud over continental dimensions. The gray particulates and colored gases commenced to darken the skies rapidly diminishing daylight. These effects would likely produce an environmental disaster akin to the "Nuclear Winter" described by many scientists, including Carl Sagan of Cornell University (Sagan et al. 1983). They predicted that ash from atomic explosions in a hypothetical "5000 Megaton war" together with soot from the resulting raging fires, would pervade the atmosphere, obscuring sunlight in both hemispheres. The results were predicted to globally shut down photosynthesis, while forcing air temperatures to plummet as much as -20 °C—even in the temperate zones. This would "refrigerate" the Earth for at least several months, resulting in atmospheric condensation and causing much of the Earth to be pelted with rain, sleet, and snow—all of which was highly radioactive.

When comparing Nuclear Winter effects with those of Chicxulub, the good news is that with Chicxulub, no radioactivity was involved. But, the bad news was that Chicxulub's explosive KE was -10,000 times greater than Sagan's postulated atomic blasts. Moreover, the mass of Chicxulub's ejected debris had to be immense—recall from Alvarez's studies that some 60 times Chicxulub's mass in Earth material was also blown into the atmosphere.

As was estimated, about 20 % of the Chicxulub ejecta (several Gigatons) of the finer soot and particulate matter remained aloft for decades, if not years. It has been suggested that global wind patterns had morphed into gigantic cyclonic weather systems that distributed the ash and gases worldwide. Moreover, the onset of darkness exacerbated both rapid cooling with the shutting down of photosynthesis to a much higher degree than that prescribed by Sagan's Nuclear Winter, especially in the Northern Hemisphere. This cooling first caused torrential rains (Fig. 5.3) and worse, in the middle and higher latitudes became intense continental-sized blizzards that deposited meters of snow per day.

Fig. 5.3 Schematic portrayal of the super storm, generated from the boiling vapors and condensing above the impact crater. The hot vapors contained high concentrations of toxic chemicals rose into the upper atmosphere and condensed, forming monstrous monsoon-like rains and snow blizzards. This contributed to global conditions causing the "Nuclear Winter"

As months turned into years, the crater bottom-melt sheet cooled and the seawater ceased to boil, eliminating one source of the Earth-shrouding toxic fog. In the lower atmosphere, precipitation helped to flush out the sulfurous toxins while the ash and soot settled out—this allowed the daytime skies to slowly brighten, clear, and warm. At the same time, the vast heat stored in the upper layers of the mid-latitude and tropical oceans, stirred by the extreme storm winds, started to mix upward replenishing more heat to the lower atmosphere. As the Earth adjusted to a rejuvenated warming spring, one last monster, however, reared its ugly head.

The Long "Greenhouse" Summer

As the heating continued, the surviving remnants of the Earth's terribly damaged biosphere suffered a final blow lasting for at least hundreds of years. Solar heating in the clearing atmosphere now accelerated and began to wildly overshoot the pre-Chicxulub atmospheric temperatures. This effect was driven by Gigatons of greenhouse gases that were deposited in the atmosphere from the impact explosion. These included: (1) carbon dioxide produced from the combustion of vaporized bottom sediments and its underlying limestone layers as well as from the vegetation of land fires, (2) sulfur oxides derived from the vaporization of the sulfate rocks blown from the crater, and (3) vaporized Gulf water from the crater explosion. These strong "Greenhouse Gases", unlike the settling denser particulates, would linger far longer in the atmosphere.

With the Earth again warming, the continental-sized snowfields melted away, causing global flooding and raised the coastal sea level several meters. Slowly, inexorably, vast areas of the temperate zone were transformed from swamps into sweltering deserts, as most of the world's lakes were evaporated dry. This over-heating of the Earth was compounded by an increase in evaporation rates over the oceans, insuring that the Earth's lower atmosphere maintained high levels of water vapor—one of the most potent greenhouse gases of all.

After a few hundred years, the acidic aerosols and the greenhouse gases gradually decomposed and were purged from the Earth's atmosphere, land, and oceans. Skies cleared as the atmosphere gradually returned to its former state before the Great Chicxulub event. This "Perfect Storm," which encompassed a chain of disasters that ravaged the Earth's land and ocean environments had mercifully passed.

We have devoted much of the Chicxulub story to describing the scenario of physical effects upon the Earth's land, ocean, and atmosphere. If there had been a complete absence of life on Earth at the time of the Chicxulub—our story could end here. But alas, life was very much present at the time. Some life forms were affected more drastically than others. As the authors are physical scientists, we offer only a brief summary of these effects upon the Earth's biota. For a broad discussion of the subject see, for example, Schulte et al.'s review (2010).

Damage to the Biosphere

The Earth's biosphere had suffered sorely from disastrous cosmic artillery. After a period of greenhouse heating, the Earth began the slow process of healing. As we have discussed previously, the data suggests that in all, some 60–80 % of life on Earth had been destroyed. As suggested in Fig. 3.6, this destruction took many forms: e.g., instant incineration, blast trauma, drowning, freezing, starvation, poisoning, acidification, suffocation, dehydration, and baking—all of these pretty grim.

The damage occurred in several stages. First, the pressure and heat waves from the initial impact explosion immediately destroyed life on land within a radius of a thousand or so kilometers from ground zero. Shortly thereafter, the atmospheric pressure wave arrived and further devastated life on land. While within the seas, incredible acoustic pressures, generated from the ocean splash and explosion, shocked the marine life. Then came the deadly heat wave emanating from the searing ejecta as it was falling back to Earth causing IR radiation to reach thousands of degrees at ground level. Although lasting only a few minutes, it mercilessly scorched plants and animals, setting off fires of land vegetation over much of the Western Hemisphere.

Within 1 or 2 hrs of impact, chaotic tsunamis raced across the seas, as we discussed in Chap. 4. Continental flooding was massive, reaching far inland. The giant breakers, moving into the Atlantic Ocean, drowned a significant fraction of land animals along the American and European coasts, and then proceeded to drown coasts on both sides of the South Atlantic Ocean. At long last, a period of deathly silence fell over most of the planet.

For the creatures surviving the first few days after the onslaught of fire, wind, and waves, further devastation took place more gradually. Within weeks to months, the residual effects of poisoning of the land and the upper ocean would have taken hold. Meanwhile, there was a period of global freezing. Then, there was a long period of atmospheric baking, followed by the subsequent extreme roller coaster climate changes that followed. Each of these phenomena decimated or eliminated entire food chains.

Paleontological evidence suggests that no other mass extinctions covering such a broad range of creatures (i.e., from foraminifera to dinosaurs) had ever been detected over the millions of years either before or after the K-Pg event. Had these events occurred within the span of human history, instead of the end of the Cretaceous Period, it would have severely disrupted civilization throughout the globe.

As we have indicated in Chap. 2, large land vertebrates including the dinosaurs, marine and flying reptiles, as well as flesh eating animals greater than ~25 kg, abruptly disappeared across the fossil record in the K-Pg boundary. This boundary also marked the pronounced loss of diverse vegetation, leafy land plants and forest communities, all of which must have had a deleterious effect upon surviving herbivorous and other life forms.

In the Oceans, fossil sediments displayed abrupt decreases in productivity and the primary producers such as zoo-and phytoplankton. Nearly all microscopic floating plankton and calcium—based foraminifera, especially those dependent on photosynthesis, died off. Whole ranges of rudists (spiral shelled mollusks), and calcareous nanofossils, disappeared over the K-Pg boundary.

Land creatures, who had lived through the initial radiation of heat and shock, survival seemed a matter of finding a place to hide from the cosmic bombardments and the extreme torrential precipitation. Alas, the ponderous dinosaurs and flying reptiles were perhaps too big and ill equipped to find safe cover or enough to eat amid these periods of rapidly changing environments.

And what of the survivors? Curiously, the fossil evidence indicates that a disproportionate number of late Cretaceous rodent-like land mammals were able to tough it out. These omnivorous nocturnal creatures were able to escape and survive the cold, wet and dark environments probably by burrowing into rocks and caves and sustained by frozen (preserved) vegetation. It also appears that the ancestors of these smaller mammals had existed before the Chicxulub impact, but that they had played only a minor role in the animal kingdom, which, until the impact, had been dominated by non-avian dinosaurs.

A recent study led by Dr. Maureen O'Leary of Stony Brook University at Long Island, NY suggests that the demise of the dinosaurs actually promoted the flourishing of these early mammal forms (O'Leary et al. 2014). One example was of a possum-like creature named "Protungulatum donnae." Its fossil characteristics anticipated all "placental" mammals, i.e., as opposed to the reptilian kingdom most of whose offspring hatched from eggs. These creatures seemed to have become dominant within only ~300,000 years (a mere "fortnight" in geologic time) after the impact extinctions. They, subsequently, became the forbearers of the some 5400 present diverse species of the mammalian kingdom (as Dr. O'Leary notes: from shrews to elephants, bats to whales, cats to dogs, and humans). There remain of course, many unanswered questions as to what propelled the rapid expansion of the new mammalian tree of life and over what time periods.

In any case, a variety of new and unique forms sprung up during the millennia's after that fateful day when Chicxulub arrived. (That so many of these small mammals had virtually replaced the mighty dinosaurs, suggested that the meek might inherit the Earth after all). An interesting question remains as to what affects this sudden removal of such a large fraction of the Earth's species had upon subsequent evolutionary processes that led to life on Earth today. This question is attended to in the following Epilogue.

References

O'Leary M et al (2014) http://www.the-scientist.com/?articles.view/articleNo/38858/title/Clocks-Versus-Rocks/. Accessed 2 Jan 2012

Sagan C, Turco R et al (1983) Nuclear winter: global consequences of multiple nuclear explosions. Science 222(4630):1283

Schulte P et al (2010) The Chicxulub asteroid impact and mass extinction at the Cretaceous-Paleogene boundary. Science 327:1214–1218

Chicxulub: The Impact and Tsunami

The Story of the Largest Known Asteroid to Hit the Earth

David Shonting and Cathy Ezrailson

© Springer International Publishing Switzerland 2017
D. Shonting, C. Ezrailson, *Chicxulub: The Impact and Tsunami*,
Springer Praxis Books, DOI 10.1007/978-3-319-39487-9

DOI 10.1007/978-3-319-39487-9_6

The original version of this book was inadvertently published with an incorrect gravity symbol $(m/s)^2$ in page no: xviii. The correct symbol is updated as m/s^2.

Chapter 1: The Orbiting Objects

The original version of this chapter was inadvertently published with an incorrect line in the footnote in page-6. The end of first sentence in the footnote has been updated as "~10 m/s every second."

Chapter 2: The Tale of Chicxulub

The original version of this book was inadvertently published with an incorrect bank name in page no 27; under the topic "Found: A Crater" lines 4 & 14 and page no 32; 3rd paragraph, bottom line. The correct bank name is updated as: Campeche Bank.

The updated online version of the original book can be found at
http://dx.doi.org/10.1007/978-3-319-39487-9

© Springer International Publishing Switzerland 2017
D. Shonting, C. Ezrailson, *Chicxulub: The Impact and Tsunami*,
Springer Praxis Books, DOI 10.1007/978-3-319-39487-9_6

Chapter 3: A Scenario for the Chicxulub Impactand Energies

The original version of this book was inadvertently published with an incorrect word in page no: 59: 14th line. The correct word is updated as: Sunday.

Chapter 4: The Chicxulub Tsunami

The original version of this book was inadvertently published with incorrect order of figure 4.10a and 4.10b in page no: 84. The correct order is updated as:

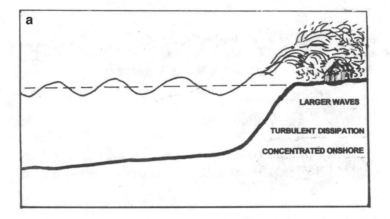

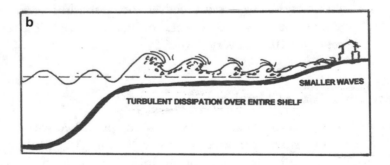

In page no: 104 (bottom line) the speed of waves spread was published with incorrect unit; the correct unit is updated as: ~200–250 m/s (400–500 knots).

Epilogue: Possible Chicxulub Effects on the Path of Human Evolution

The original version of this book was inadvertently published with an incorrect bank name in page no 120 (paragraph 4 & line 7). The correct bank name is updated as: Campeche.

Epilogue: Possible Chicxulub Effects on the Path of Human Evolution

We have portrayed the story of the great Chicxulub impact with Earth and briefly commented on the resulting damage to its plant and animal life. At this period 65.5 million years ago, it seems worthwhile to ask, "What were the possible effects upon the fate of human evolution."

A variety of possible evolutionary histories inevitably produce a host of "what-if questions." One question might be, what if the asteroid, instead of splashing into the ancient Gulf adjacent to the Atlantic Ocean, had crashed into the center of a continent? It turns out that the effects upon the biologic environments would be far different. First, a huge volume of seawater would be missing from the ejecta, producing a much smaller atmospheric impact plume. With a smaller plume, the degree of decreasing sunlight, the amount of Earth cooling, and the shutting down of photosynthesis would be much less. Furthermore, with no superheated seawater to react with unearthed sulfates and oxides, the toxicity of the plume would also reduced. Moreover, the absence of a boiling crater caldron would have eliminated another source of toxic vapors feeding the atmospheric cloud. Thus, the volume of the subsequent acid rains, compared to that of an ocean impact cloud would have been much less. Also, of course, there would have been no tsunami. In short, the Earth's biosphere would have been far less adversely affected by a land impact of Chicxulub. As we have noted, because the oceans have occupied approximately three quarters of the Earth's surface for over the past 100 million years, the odds were ~3:1 for the Earth—bound Chicxulub to make an ocean impact. And, the odds had it!

The original version of this chapter was revised. The erratum to this chapter is available at: DOI 10.1007/978-3-319-39487-9_6

© Springer International Publishing Switzerland 2017
D. Shonting, C. Ezrailson, *Chicxulub: The Impact and Tsunami*,
Springer Praxis Books, DOI 10.1007/978-3-319-39487-9

Perhaps, a more profound question might be: What, if any, were the effects of the Chicxulub event upon the ultimate path of mammalian (and later, human) evolution over the millions of years since the impact?

In a fascinating book, "Wonderful Life" (1990) by Steven J. Gould, the late Harvard biologist, discusses the history of life and how changes in the sequence of events can alter the paths of evolution of plants and animals. He describes the discovery in 1909 of the small, fossilized ancient seabed high in the Canadian Rockies—the famous Burgess Shale. This is a 530 million year-old remarkably preserved layer of Cambrian-era marine fossils with more varieties of sea life than is found in our modern oceans. Most importantly, this miraculous collection yielded an entirely new spectrum of fossils of "soft bodied" marine creatures having many strange forms found nowhere else. Such soft parts are rarely preserved in the fossil world since the effects of scavenging and decay usually eliminates them.

How did such preservation occur? One theory speculates that a sedimentary bed had formed at the bottom of a shallow sea, which by a stroke of chance became isolated from the greater world oceans (see Fig. E.1). This inland sea environment had to have been isolated long enough to evolve unique life forms within their own microcosm and with the passage of millennia, the marine animal forms lived out their evolutionary paths—and became extinct.

The bottom environment, where the dead creatures had collected, must have consisted of a chemically reducing medium—devoid of oxygen, as would occur in a stagnant basin. This could have enhanced the preservation of the soft parts of the animals. The present shale sediment formation suggests that bottom animal gravesites were traumatically shifted or buried deeper by an earthquake or turbidity current. Continued sedimentation above these sites, taking place over eons, caused the layers to consolidate, allowing the Burgess Faunal soft parts to be fossilized along with their bony parts.

Since the long periods of isolation had allowed many of the Burgess creatures to become extinct, their forms would be lost forever, were it not for the discovery of this shale quarry. Gould draws from the idea of how happenstance and fate can determine life's trajectories and milestones. From his study of this shale formation, Gould poses the question of the meaning of the "tree of life." Is it a ladder of progress governed only by the survival of the fittest, promoting more refined and superior creatures? Or, is it more of a lottery, as in Chaos theory, where small random perturbations cause changes to the path of life's evolution, which then can magnify and spread out in many directions.

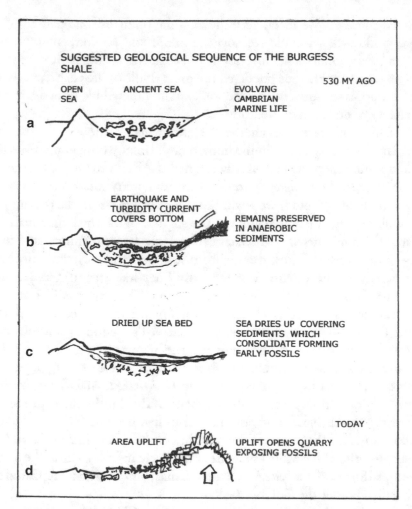

Fig. E.1 Suggested sequence for the formation of the Burgess Shale. An ancient Cambrian sea **(a)** with an abundance of marine life, becomes isolated from the open ocean. A catastrophic event **(b)** covers the anaerobic sediments and the pre-served remains of the sea life. After eons covering sediments consolidate, forming ancient fossils **(c)**. Mountain building and uplift opens up the shale quarry **(d)**

And where many of the branches of this tree, being pure accidents of fate are doomed to failure?[1]

---

[1] It might have been, that while Professor Gould was writing his book, just ten or so city blocks east of Harvard Square, M.I.T. meteorology Professor Edward Lorenz, the father of Chaos Theory, was perhaps still playing back his computer programs and questioning how "The flap of a butterfly's Wings in Brazil Set off a Tornado in Texas," the subject of a paper he delivered to the American Association for the Advancement of Science in 1972.

Gould muses; if we metaphorically at some point in time play the "tape of life" backwards we could see how the minor and random environmental changes create whole new universes of life creatures far different than in the previous pathway. He thus pondered the great riddle of life's existence, which when extrapolated forward in time, suggests the incredible odds against the probability of our human evolution.

The Chicxulub event was another instance where a throw of the dice may have played a vital role, say, in the human evolutionary outcome. At the time of Chicxulub, dinosaurs had already roamed the Earth for at least some 200 million years so their longevity to that time was pretty much a success story. Now, if (following Gould) we rewind the "tape of life" from the present back 65.5 million years, we note that it was the survival of the small mammals who replaced the dinosaurs and who later evolved into larger mammals and much later—into ourselves. Thus, we might muse about which events contributed most heavily to the evolution of the early Lucy-like creatures and eventually to us (Homo sapiens). Key questions are: What chances of fate could have intervened if Chicxulub had had a slightly different trajectory and never hit the Earth, removing the existence of the K-Pg boundary and obviating the catastrophic demise of the dinosaurs and many other of the life forms? Moreover, how different would today's scenario of life be on our planet?

Now, fast-forward 65.5 million years to the present. Alas, today is reality and the Chicxulub impact actually happened. The continents kept inexorably moving about. North and South America had become joined, producing Central America, The return of the tiny foraminifera's calcium shells and carbonate precipitates have filled sections of the ancient Gulf to form the mighty coastal platforms of the Great Bahamas Banks, Florida, and the Campeche Bank which border the Gulf of Mexico.

And, far beneath this Bank lie the vestiges of the Chicxulub, its massive crater structures are at rest, frozen in time; part beneath the Yucatan peninsula and part deep under the shallow Bank. The multiple crater rims are covered with kilometers of limestone—fit for a great Pharaoh's tomb. Were we to re-visit this present site from high in the sky, far above the mighty hidden monolith, the inhabitants of the little fishing village of Chicxulub might be observed swinging in their hammocks as they gaze out over the serene gin-clear water of the Campeche Bay, or plying over it in their small fishing boats (see Fig. E.2) going about their daily lives. Chances are that most of them could never imagine the truly earthshaking scenario that had occurred at that very spot beneath them so many millions of years ago, when the Mighty Chicxulub object fell from the sky.

Fig. E.2 The small fishing boat at Port Chicxulub rests serenely in the gin clear waters far above the ancient crater Photo permission by Deidre Mize (from Blog: http://www.awolamericans.com/pangas-of-playa-del-carmen)

That event would indeed be the most ancient of histories—much too early to imagine. Indeed, one might wonder that if the Chicxulub event had been instead, a narrow miss, what sort of creatures would be basking in the sunshine on the tropical beach near the village of Chicxulub. Might they be intelligent lizards wearing sombreros, puffing on a marihuana cigarette, while manipulating their iPhones?

Reference

Gould SJ (1990) The burgess shale and the nature of history. In: Gould SJ (ed) Wonderful life. W. W. Norton, New York, p 347, Illustrated

Additional Resources

The following web resources may help the interested reader to pursue in greater depth the study of the sciences addressed in this book and beyond.

Chapter 1

Chapman CR, Morrison D (1994) Impacts on the Earth by asteroids and comets: assessing the hazard. Nature 367(6458):33–40

The Hustle and Bustle of Our Solar System (NASA) http://www.nasa.gov/multimedia/imagegallery/image_feature_2319.html

The Near Earth Orbit Program (NASA) http://neo.jpl.nasa.gov/

Chapter 2

Planetary disasters: It could happen one night. http://www.nature.com/news/planetary-disasters-it-could-happen-one-night-1.12174

The Geological Time Scale (The Geological Society of America). http://www.geosociety.org/science/timescale/timescl.pdf

Hildebrand AR et al (1995) Size and structure of the Chicxulub crater revealed by horizontal gravity gradients and cenotes. Nature 376(6539):415–417

Pilkington M, Hildebrand AR (2000) Three-dimensional magnetic imaging of the Chicxulub crater. J Geophys Res 105(B10):23479–23491

Morgan J et al (1997) Size and morphology of the Chicxulub impact crater. Nature 390(6659):472–476

© Springer International Publishing Switzerland 2017

D. Shonting, C. Ezrailson, *Chicxulub: The Impact and Tsunami*,

Springer Praxis Books, DOI 10.1007/978-3-319-39487-9

Morgan J, Warner M (1999) Chicxulub: the third dimension of a multi-ring impact basin. Geology 27(5):407–410

Chapter 3

MITVideo (2015) Nuclear explosions and the effects of shock waves. MIT Tech TV, Edgerton Center. http://video.mit.edu/watch/nuclear-explosions-and-the-effect-of-shock-waves-2620/. Accessed 26 Dec 2015

Old Faithful Geyser Streaming Webcam (2015) http://www.nps.gov/features/yell/webcam/oldFaithfulStreaming.html. Accessed 27 Dec 2015

Oskin B (2014) LiveScience. http://www.livescience.com/23387-mariana-trench.html. Accessed 29 Oct 2014

Planetary Science Institute (2014) Impact craters virtual tours. http://www.psi.edu/epo/explorecraters/virtualtours.htm. Accessed 27 Dec 2015

The Independent (2006) How Krakatoa made the biggest bang. http://www.independent.co.uk/news/science/how-krakatoa-made-the-biggest-bang-5336165.html. Accessed 10 Oct 2015

Zhang M (2013) Photos from the world's first underwater nuclear explosion, petapixel. http://petapixel.com/2013/02/18/photos-from-the-worlds-rst-underwater-nuclear-explosion/. Accessed 12 Nov 2013

Chapter 4

Hosaka T (2011) How one Japanese village defied the tsunami. Today News. Accessed 20 Jun 2012

International tsunami information center (2015) How do earthquakes generate tsunamis? http://itic.ioc-unesco.org/index.php?option=com_content&view=article&id=1158&Itemid=2026. Accessed 2 Dec 2015

National weather service Jetstream (2015). Online school for weather: Tsunami generation. http://www.srh.noaa.gov/jetstream/tsunami/generation.htm. Accessed 9 Sep 2015

Stewart R (2008) Introduction to physical oceanography. Open Textbook Library. https://open.umn.edu/opentextbooks/BookDetail.aspx?bookId=20. Accessed 28 Dec 2015

Chapter 5

Hsu J (2010) Asteroid strike could force humans into twilight existence, LiveScience 249 26 Oct 2010. http://www.livescience.com/8825-asteroid-strike-force-humans-250 twilight-existence.html. Accessed 3 Dec 2015

Smith L (2014) What killed the dinosaurs? Post-Asteroid collision "Impact Winter" caused intense global cooling, IBT 13 May 2014. Accessed 14 Dec 2015

Printed in the United States
By Bookmasters